Static Analysis of Determinate and Indeterminate Structures

Static Analysis of Determinate and Indeterminate Structures

Kenneth Derucher, Ph.D., P.E.
Professor and Former Dean of College of Engineering
Department of Civil Engineering
California State University, Chico

Chandrasekhar Putcha, Ph.D., F. ASCE
Professor
Department of Civil and Environmental Engineering
California State University, Fullerton

Uksun Kim, Ph.D., P.E.
Professor
Department of Civil and Environmental Engineering
California State University, Fullerton

Hota V.S. GangaRao, Ph.D., P.E.
Wadsworth Professor
Wadsworth Department of Civil and Environmental Engineering
Statler College of Engineering
West Virginia University, Morgantown

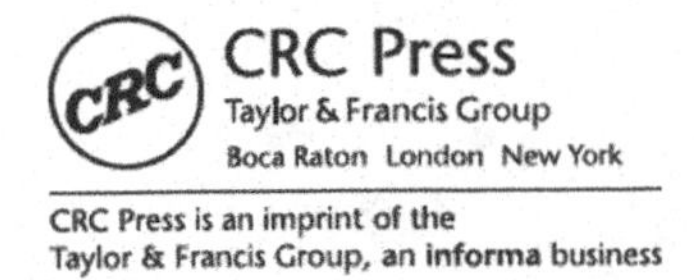

CRC Press is an imprint of the
Taylor & Francis Group, an informa business

First edition published 2022
by CRC Press
6000 Broken Sound Parkway NW, Suite 300, Boca Raton, FL 33487-2742

and by CRC Press
2 Park Square, Milton Park, Abingdon, Oxon, OX14 4RN

CRC Press is an imprint of Taylor & Francis Group, LLC

Library of Congress Cataloging-in-Publication Data
Names: Derucher, Kenneth N., 1949– author. | Putcha, Chandrasekhar, author. |
Kim, Uksun, author. | GangaRao, Hota V. S., author.
Title: Static analysis of determinate and indeterminate structures /
Identifiers: LCCN 2021031416 (print) | LCCN 2021031417 (ebook) |
ISBN 9781032159829 (hbk) | ISBN 9781032159904 (pbk) |
ISBN 9781003246633 (ebk)
LC record available at https://lccn.loc.gov/2021031416
LC ebook record available at https://lccn.loc.gov/2021031417

ISBN: 978-1-03-215982-9 (hbk)
ISBN: 978-1-03-215990-4 (pbk)
ISBN: 978-1-00-324663-3 (ebk)

DOI: 10.1201/9781003246633

Typeset in Times
by Newgen Publishing UK

Contents

PART II Analysis of Statically Indeterminate Structures

Preface

The title of this book is *Static Analysis of Determinate and Indeterminate Structures*, and not "Structural Analysis". The title is a bit unique as we want to specify that this is a static analysis (as opposed to dynamic analysis) of both determinate and indeterminate structures. Many textbooks have been written on structural analysis over the past several years with a twofold composition. They essentially deal with analysis of statically determinate structures followed by analysis of statically indeterminate structures using the force method, displacement methods (classical methods such as moment distribution), and the stiffness method. Thus, the material covered in existing textbooks on structural analysis contains more than what is necessary to learn indeterminate structural analysis. As a result, these books become bulky and all their material cannot, and need not, be covered in a single course on indeterminate structural analysis. Moreover, these books rarely include an as-needed discussion of the unit load method, which is arguably the best method to calculate deflections when solving problems by the force method. Hence, the authors set out to create this book.

The first part of this book essentially deals with the analysis of determinate structures (in a total of five chapters). The second part of the book deals with the analysis of indeterminate structures by force method, displacement methods stiffness method (in a total of seven chapters).

The first chapter deals with solving reactions using equations of force. The second chapter deals with deflections of statically determinate beams and frames. The third chapter deals with deflections of statically determinate trusses. The fourth chapter deals with shear force and bending moment diagrams for beams. The fifth chapter deals with influence lines for statically determinate structures. The emphasis is mainly on beams.

The sixth chapter deals with application of the force method to analysis of beam, frame, and truss structures. The unit load method is discussed with reference to the analysis of statically indeterminate structures. A few examples are discussed to illustrate these concepts. The seventh and eighth chapters deal with analysis of indeterminate structures by displacement methods. In Chapter 7, the concepts of slope-deflection method are developed and applied to beam and frame structures. The eighth chapter deals with developments of concepts of the moment distribution method. These concepts are then applied to beam and frame structures. The ninth chapter develops the concepts of the stiffness method. These are subsequently applied to beam structures. The tenth and eleventh chapters deal with application of the stiffness method to frame and truss structures, respectively. The 12th chapter deals with approximate analysis of indeterminate structures. Throughout the book, illustrative examples are discussed under each method. The intent is to cover as much material as is needed conceptually with minimal and yet sufficient examples so that students can understand indeterminate structural analysis methods without being overwhelmed. In this way, the book is kept less bulky compared to existing books on structural analysis. In addition, keeping the textbook concise will reduce the price far below that of existing textbooks, saving money for students. We believe that this will be a big selling point because the amount of material covered is not compromised in covering the material in a concise manner. This is in addition to the fact that this book is written by four full professors of Civil Engineering who have had vast experience in teaching and research in the area of structural analysis.

It is hoped that this experience is reflected in the write-up of this book so that it serves our twofold objective. First, we hope that the instructor following this book as a textbook for his/her course on determinate or indeterminate structural analysis feels that all the required material is indeed covered in this textbook. Second, we hope that students taking this course will find the book and material covered inside easy to understand.

The authors are thankful to Mr. Kyle Anderson and Mr. AnhDuong Le, former graduate students in the Department of Civil and Environmental Engineering at California State University, Fullerton, for going through the manuscript and making constructive comments in Part II of the book dealing with indeterminate structural analysis. We also appreciate the editing work done by Mr. Alexander Motzny, undergraduate student in the Department of Civil and Environmental Engineering at California State University, Fullerton.

Kenneth Derucher
Chandrasekhar Putcha
Uksun Kim
Hota V.S. GangaRao

Author Biographies

Dr. Kenneth Derucher is a Professor of Civil Engineering at California State University, Chico (Chico State) and has served as the Dean of the College of Engineering, Computer Science and Construction Management. Dr. Derucher is a teacher, researcher, consultant, fundraiser, and author of a number of textbooks and various publications. He is a professional engineer who has served on many legal cases as an expert witness in the engineering field. During his career, he has received many accolades for his achievements.

Dr. Chandrasekhar Putcha is a Professor of Civil and Environmental Engineering at California State University, Fullerton, since 1981. Before that, he worked as a Research Assistant Professor at West Virginia University, Morgantown. His research areas of interest are reliability, risk analysis, and optimization. He has published more than 180 research papers in refereed journals and conferences. He is a fellow of the American Society of Civil Engineers. He did consulting for federal agencies such as National Aeronautics and Space Administration (NASA), Navy, US Army Corps of Engineers, and Air Force, as well as leading private organizations such as Boeing and Northrop Grumman Corporation (NGC).

Dr. Uksun Kim is a Professor of Civil and Environmental Engineering at California State University, Fullerton and served as the Department Chair from 2012 to 2018. His research interests include seismic design of building systems with steel joist girders, partially restrained connections, and seismic rehabilitation of prestressed building systems. He is a licensed professional engineer and a Leadership in Energy and Environmental Design Accredited Professionals (LEED AP).

Dr. Hota GangaRao, after joining West Virginia University in 1969, attained the rank of Maurice and JoAnn Wadsworth Distinguished Professor in the Department of Civil and Environmental Engineering, Statler College of Engineering, and became a fellow of American Society of Civil Engineers (ASCE) and Structural Engineering Institute (SEI). Dr. GangaRao has been directing the Constructed Facilities Center since 1988, and the Center for Integration of Composites into Infrastructure (CICI), both co-sponsored by the National Science Foundation – Industry-University Cooperative Research Centers (IUCRC). He has been advancing fiber reinforced polymer (FRP) composites for infrastructure implementation, in hydraulic structures jointly with United States Army Corps of Engineers (USACE), naval vessels, utility poles, high-pressure pipes, sheet piling, and others. He chairs PHMSA's WG191 on hydraulic structures. Dr. GangaRao has published over 400 technical papers in refereed journals and proceedings, in addition to textbooks and book chapters. He has received 15 patents and many national awards. His accomplishments have been covered by CNN, ABC News, KDKA-Pittsburgh, WV-PBS, and others.

Introduction

This book deals with static analysis of both determinate and indeterminate structures. In determinate structural analysis, students learn the basic concepts of finding reactions, shear force and bending moment diagrams using force and moment equilibrium equations, and calculation of deflection of beams, frames and trusses using various methods, as well as influence lines for statically determinate structures. The book systematically covers that in detail. After that, indeterminate structural analysis is discussed. There are three basics types of methods used for analyzing indeterminate structures. They are:

1. Force method (method of consistent deformation)
2. Displacement methods (slope-deflection and moment distribution)
3. Stiffness method

General Idea about These Methods

The force method of analysis is an approach in which the reaction forces are found directly for a given statically indeterminate structure. These forces are found using compatibility requirements. This method will be discussed with more detail in Chapter 6.

The displacement methods use equilibrium requirements in which the displacements are solved for and are then used to find the forces through force-displacement equations. More on these methods can be found in Chapters 7 and 8.

The stiffness method is also considered a displacement method because the unknowns are displacements; however, the forces and displacements are solved for directly. In this book, it will be considered separately due to procedural differences from the other displacement methods. The stiffness method is very powerful, versatile, and commonly used. This method will be discussed in Chapters 9, 10, and 11. Chapter 12 deals with approximate analysis of indeterminate structures.

Part I

Analysis of Statically Determinate Structures

1 Solving Reactions Using Equations of Force

1.1 EQUATIONS OF EQUILIBRIUM

When solving for reactions in a beam, one of the most important concepts to remember is that the member (and the structure as a whole) is in equilibrium. This means that every force and moment acting on the structure is balanced by another, and the entire system remains at rest. This state of rest, or equilibrium, can be described with the following three equations of equilibrium.

$$\Sigma F_x = 0 \quad \Sigma F_y = 0 \quad \Sigma M_O = 0 \tag{1.1}$$

where F_x represents forces along the horizontal or x-axis, F_y represents forces along the vertical or y-axis, and M_O represents the sum of the moments around any point on the beam. Since these three equations can be assumed to be true for all problems, they are used in solving essentially all reactions in a structure.

1.2 DETERMINACY OF A BEAM

One limiting factor with the equations of equilibrium is that they can only solve up to three unknowns in a problem that is identified in a single plane. When a problem can be solved using only these equations, it is referred to as *statically determinate structure*. If a structure has four or more unknowns, the problem becomes impossible to solve solely using equations of equilibrium and can be called *statically indeterminate*.

The simplest way to determine if a structure is statically determinate or indeterminate is to draw the free body diagram of it, and then determine the number of reactions that need to be solved for. Several examples of this process are presented below.

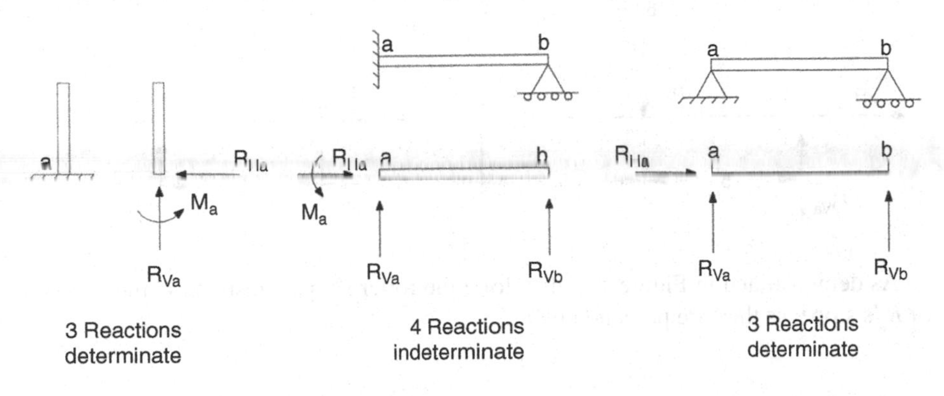

DOI: 10.1201/9781003246633-2

There are other factors that can affect whether a beam is statically determinate or indeterminate, such as hinges or shear releases. However, these will be covered in a later chapter.

1.3 EXAMPLES

EXAMPLE 1.3.1

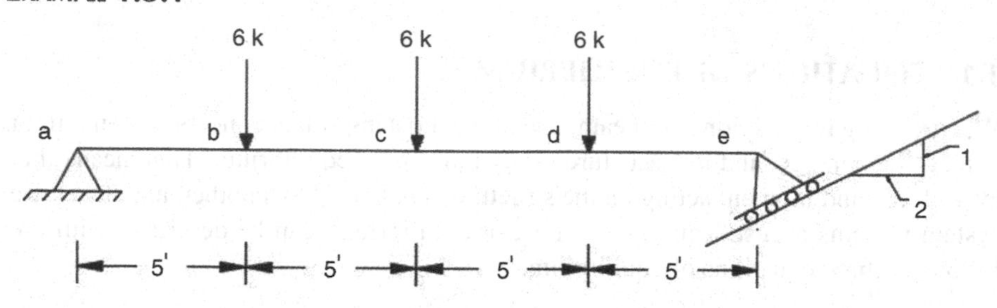

FIGURE 1.1 Beam with Hinge at Support 'a' and Roller with 1 in 2 Slope at 'e'.

In this example (Figure 1.1), joint *e* is placed on an inclined slope, where the angle of the slope is given. The first step is to break down the problem into a free body diagram.

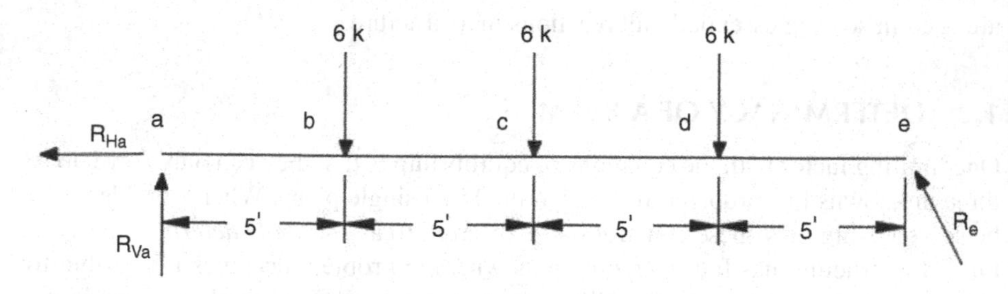

Since the equilibrium equations (equation 1.1) are only applicable in the x and y axis, R_e must be broken down and replaced by a horizontal and vertical component.

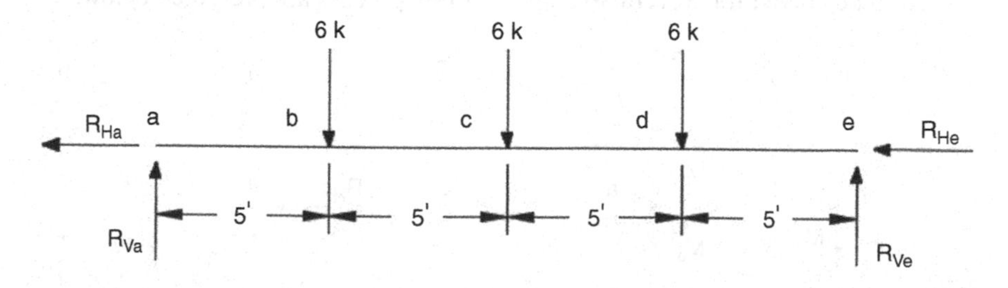

As demonstrated in Figure 1.2, the slope the roller support rests on is the inverse of R_e's slope, as they are perpendicular.

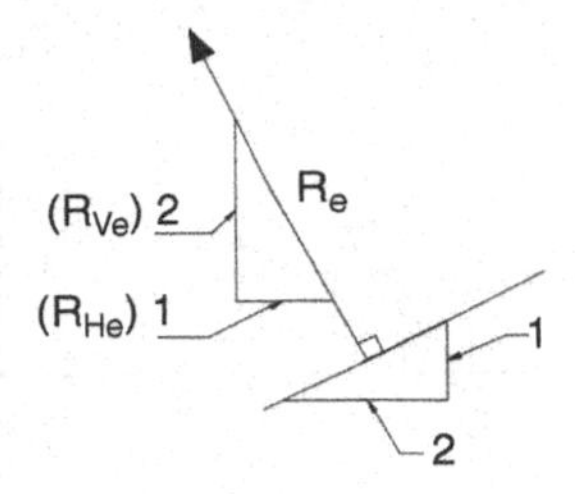

FIGURE 1.2 Reaction Re is perpendicular to the support slope.

The slope of R_e also determines the relationship of R_{Ve} and R_{He}. Therefore, since the ratio between R_{Ve} and R_{He} is 2:1, the equation can be written as

$$R_{Ve} = 2R_{He} \tag{1.2}$$

Now, the only thing that needs to be done is to write out the remaining equilibrium equations and solve the problem.

$$\begin{aligned} \Sigma M_a &= 0 = 6(5)+6(10)+6(15)-R_{Ve}(20) \\ R_{Ve} &= 9k \end{aligned} \tag{1.3}$$

Insert R_{Ve} into equation (1.3).

$$\begin{aligned} 9 &= 2R_{He} \qquad R_{He} = 4.5k \\ \Sigma V &= 0 = -R_{Va} + 6 + 6 + 6 - R_{Ve} \\ R_{Va} &= 18 - R_{Ve} = 9k \\ R_{Va} &= 9k \end{aligned} \tag{1.4}$$

$$\begin{aligned} \Sigma H &= 0 = R_{Ha} + R_{He} \\ R_{Ha} &= -R_{He} \\ R_{Ha} &= -4.5k \end{aligned} \tag{1.5}$$

To check if all of the values are correct, the sum of moments at *c* can be solved.

$$\begin{aligned} \Sigma M_C &= 0 = 10R_{Va} - 5(6) + 5(6) - 10R_{Ve} \\ &= 10(9) - 5(6) + 5(6) - 10(9) \\ &= 0 \checkmark \end{aligned}$$

EXAMPLE 1.3.2

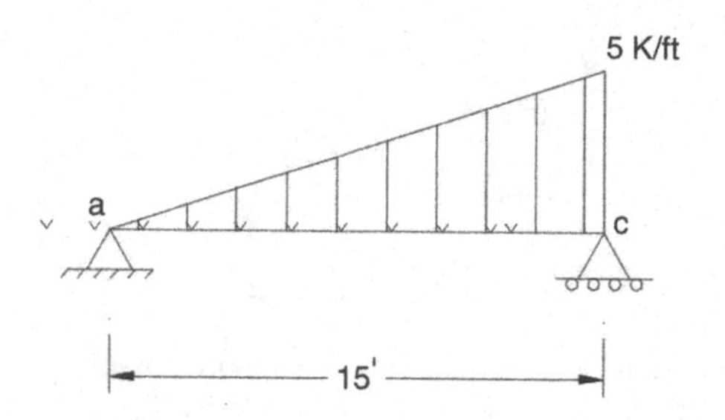

FIGURE 1.3 Simply supported beam under triangular distributed load.

The first step in solving a problem (Figure 1.3) with a distributed load is to convert the distributed load into a point load somewhere on the beam. The amount of force on the point load is equal to the area of the distributed load including load intensity. In this case, the area would be (5)×(15)/2 or 37.5 k.

The location of this load is at the geometric center of the loading shape. In this case, the beam has a triangular distributed load so the point load would be at the geometric center of the triangle, which is 1/3 of the total length from the larger end.

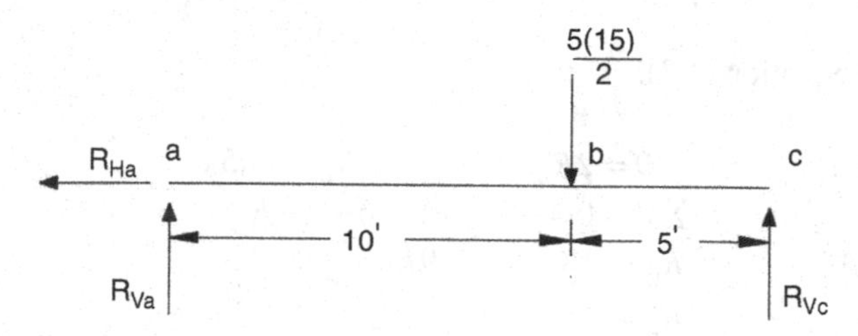

FIGURE 1.4 Simply supported beam under concentrated load at 10' from "a".

Then, after replacing the distributed load (as shown in Figure 1.4), the problem can be solved using the three equations of equilibrium.

$$\Sigma M_a = 0 = \frac{5(15)}{2}(10) - R_{Vc}(15) \qquad (1.6)$$
$$R_{Vc} = 25k$$

$$\Sigma V = 0 = -R_{Va} + \frac{5(15)}{2} - R_{Vc}$$
$$R_{Va} = 37.5 - R_{Vc} \qquad (1.7)$$
$$R_{Va} = 12.5k$$

$$\Sigma H = 0 = -R_{Ha}$$
$$R_{Ha} = 0 \qquad (1.8)$$

To check if all of the values are correct (satisfying equilibrium), the sum of moments at b can be verified to be zero.

$$\begin{aligned}\Sigma M_b = 0 &= 10R_{Va} - 5R_{Vb} \\ &= 10(12.5) - 5(25) \\ &= 0\ \checkmark.\end{aligned}$$

Example 1.3.3

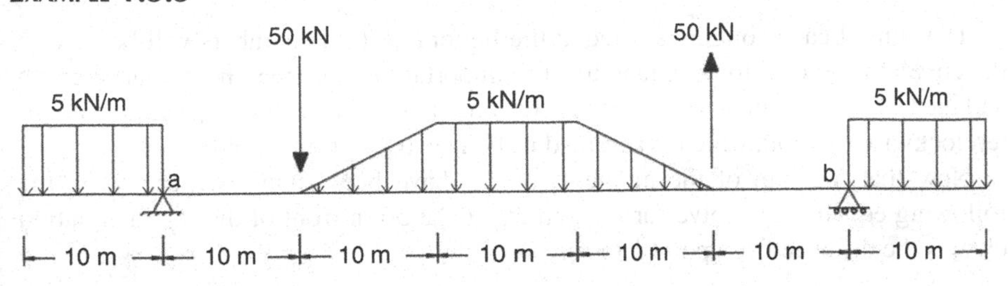

FIGURE 1.5 Simply supported beam with overhangs under varying loads.

Similar to the beginning of Example 1.3.2, the first step in this problem (Figure 1.5) is to convert all of the distributed loads into pseudo point loads on the beam.

It is possible to divide the trapezoidal distributed load located in the center of the beam into a single point load, or into three separate point loads. Both of the methods will result in the same answer. In this example, the trapezoid will be replaced with a single point load (Figure 1.6).

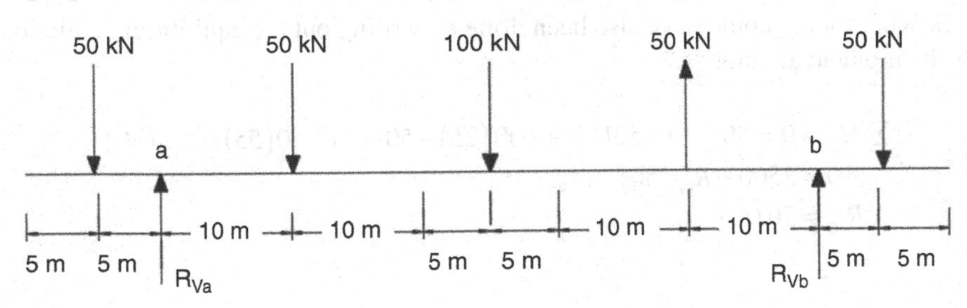

FIGURE 1.6 Single point load equivalency of loads in Figure 1.5.

Next, since there are no horizontal loads acting on the beam, R_{Ha} must be equal to zero.

$$R_{Ha} = 0$$

In some cases where there are many different vertical forces acting on a beam, a table can be used to help solve the problem faster. To make such a table, a point is first picked somewhere on the beam to find the moments from. In this example, point "a" will be selected to develop the moment equilibrium equation. Then, all the forces, the length the forces are from the point (i.e., the arm length), and the resulting moments are all added to the table (as in Figure 1.7).

Force (kN)	Arm Length (m)	Ma (kN-m)
50	5	–250
50	10	500
100	25	2500
50	40	–2000
50	55	2750
300	–	3500

FIGURE 1.7 Force & lever arm (length) and moment.

The sum of each column is listed at the bottom, as these numbers will be used in the equations (except for arm length). It is important to note that the first arm length and the fourth force are negative. This is because both of these forces produce a counterclockwise rotation, which is assumed to be negative for this problem.

Now that the sum of the moments at "a" have been found, we can write the following equations to solve for R_{Vb} and R_{Va}. (The 50 in front of the R_{Vb} in equation (1.9) is the distance from point *b* to *a*).

$$50R_{Vb} = 3500 \tag{1.9}$$

$$\begin{aligned} R_{Vb} &= 70kN \\ \Sigma V &= 0 = 300 - R_{Vb} - R_{Va} \\ 0 &= 300 - 70 - R_{Va} \\ R_{Va} &= 230kN \end{aligned} \tag{1.10}$$

Note:

Solving for R_{Vb} could have also been done by writing out the equilibrium equation for the moment at point "a."

$$\begin{aligned} \Sigma M_a &= 0 = 50(-5) + 50(10) + 100(25) - 50(40) + 50(55) - R_{Vb}(50) \\ 0 &= 3500 - R_{Vb}(50) \\ R_{Vb} &= 70kN \checkmark \end{aligned}$$

EXAMPLE 1.3.4

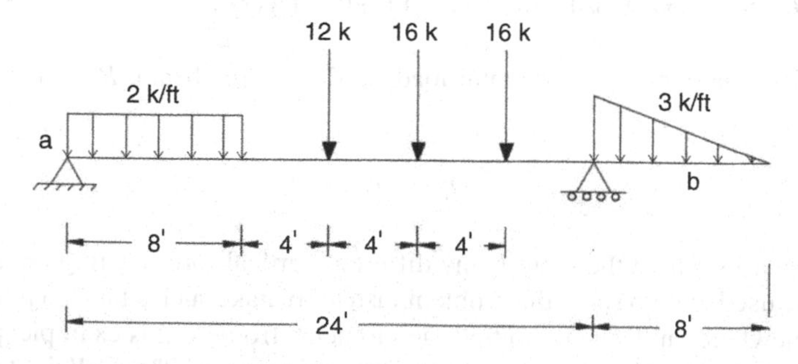

FIGURE 1.8 Simply supported beam with overhang on right side.

Similar to Example 1.3.3, this problem, shown in Figure 1.8, can be completed by using a table of moments. Since there are no horizontal forces acting on the beam, R_{Ha} is equal to zero.

$$R_{Ha} = 0k$$

The next step is then to replace both the distributed loads with corresponding point loads (Figure 1.9).

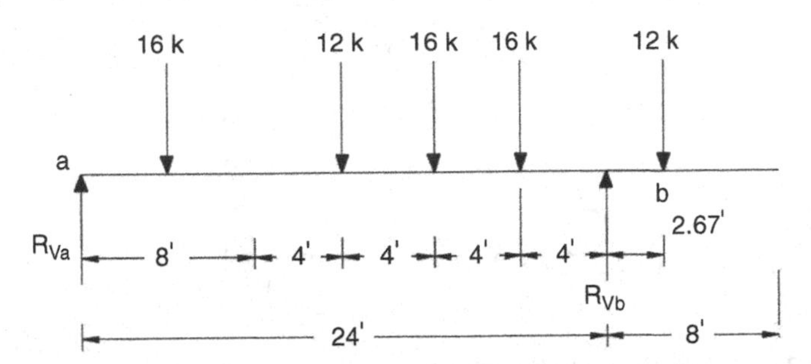

FIGURE 1.9 Simply supported beam with overhang to the right of support 'b'.

The next step is to then make a table (Figure 1.10) with the forces, arm lengths, and resulting moments (using *a* as the point of rotation).

Force (k)	Arm Length (ft)	Ma (k-ft)
16	4	64
12	12	144
16	16	256
16	20	320
12	26.67	320
72	–	1104

FIGURE 1.10 Equivalent forces as given in Figure 1.9 and corresponding lever arms and moments.

After the table has been completed, two equations can be written to find both R_{Vb} and R_{Va}.

$$\begin{aligned} 24R_{Vb} &= 1104 \\ R_{Vb} &= 46k \end{aligned} \tag{1.11}$$

$$\begin{aligned} \Sigma V = 0 &= 72 - R_{Vb} - R_{Va} \\ 0 &= 72 - 46 - R_{Va} \\ R_{Va} &= 26k \end{aligned} \tag{1.12}$$

Note:

R_{Vb} can be solved without using the table with the following equilibrium equation for the moment at "a."

$$\Sigma M_a = 0 = 16(4) + 12(12) + 16(16) - 16(20) + 12(26.67) - R_{Vb}(24)$$
$$0 = 1104 - R_{Vb}(24)$$
$$R_{Vb} = 46\ k\checkmark$$

PROBLEMS FOR CHAPTER 1

Problem 1.1 Find reactions at all supports.

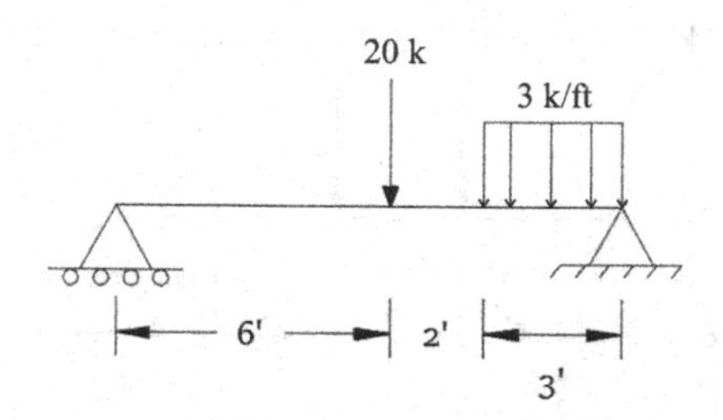

Problem 1.2 Find reactions at all supports.

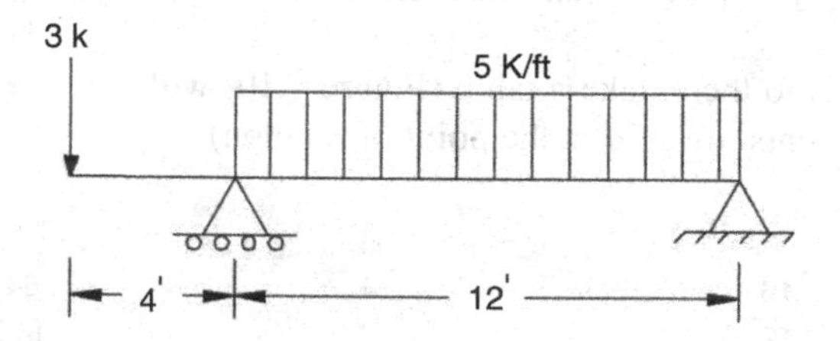

Problem 1.3 Find reactions at all supports.

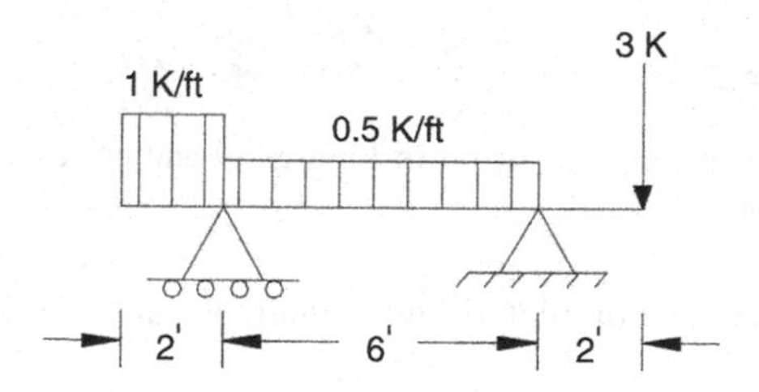

Problem 1.4 Find reactions at all supports.

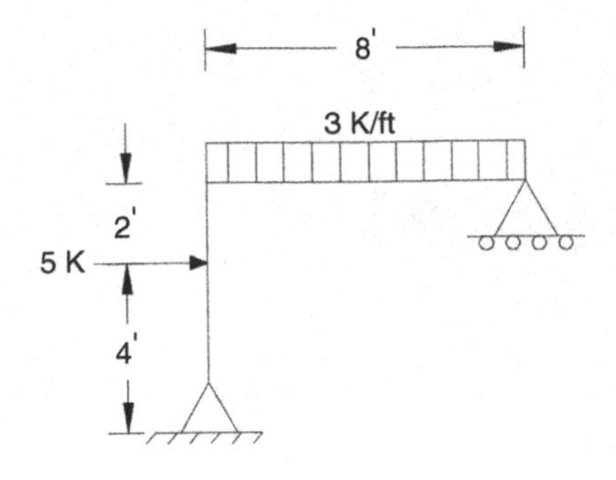

Problem 1.5 Find reactions at all supports.

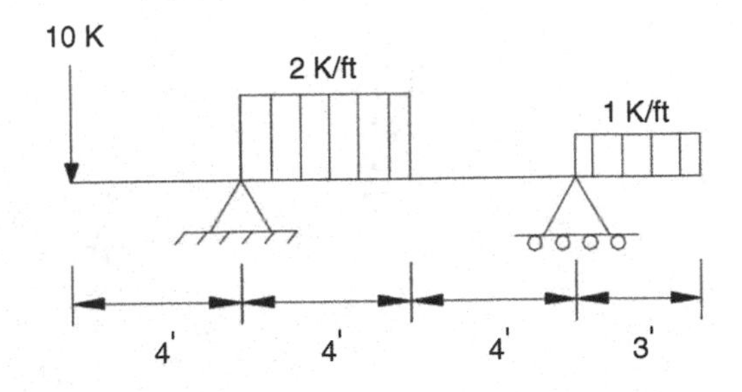

Problem 1.6 Find reactions at all supports.

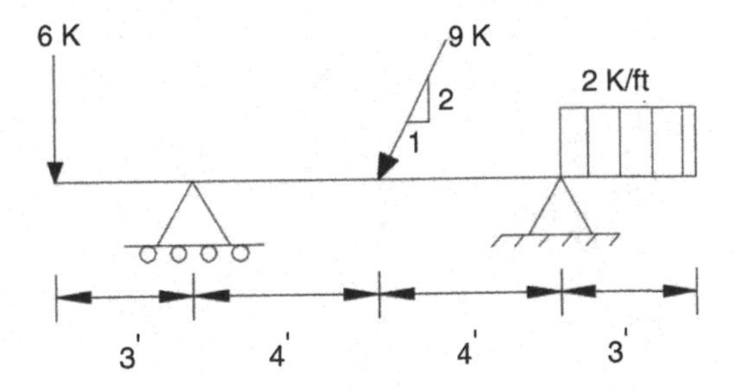

2 Deflection of Statically Determinate Beams and Frames

This chapter on deflections is very essential for both determinate and indeterminate structures. In determinate structures, the main aim is to find deflection. In indeterminate structures, the deflections are needed to eventually find the reactions at every support.

There are mainly four methods used for the calculation of deflections. These are listed below.

1. Double integration method
2. Moment area method
3. Conjugate beam method
4. Unit load method

2.1 DOUBLE INTEGRATION METHOD

The differential equation for a flexural member subjected to pure (Figure 2.1) bending is given as:

$$\frac{d^2y}{dx^2} = \frac{Mx}{EI}$$

where M, E, and I are the moment, modulus of elasticity, and the moment of inertia at that section, respectively.

This chapter deals with finding deflection y at any given distance x for the flexural member shown in Figure 2.1.

DOI: 10.1201/9781003246633-3

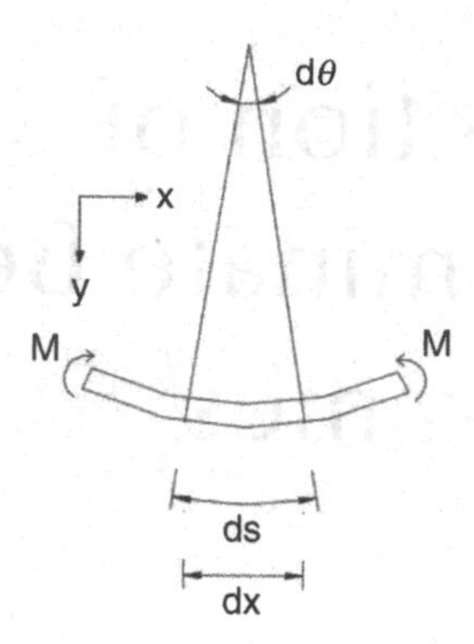

FIGURE 2.1 Deflected shape of a beam under bending causing tension at bottom.

Looking at the equation, it can be seen that integrating the left-hand side once gives *dy/dx*, which represents slope. Integrating a second time gives *y*, which is the deflection. Hence, the name of this method is double integration method.

The following examples illustrate the application of the double integration method to beams.

Example 2.1.1

Derive a general expression for deflection for a cantilever beam, subjected to a concentrated load as shown in Figure 2. 2. Use the double integration method. Then, find the deflection at the free end of the beam. The flexural rigidity *EI* is constant. Choose the fixed support as the origin.

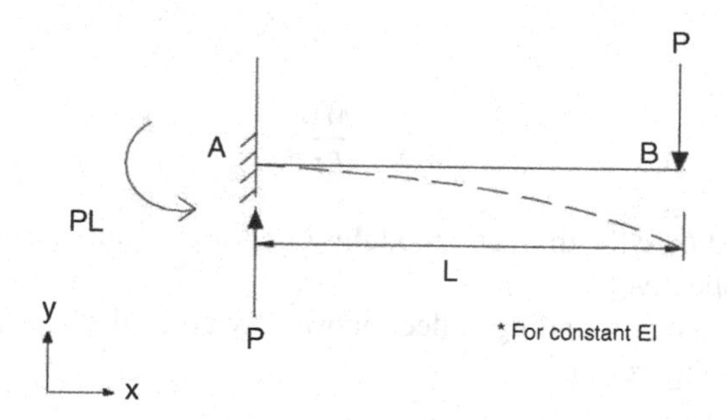

FIGURE 2.2 Deflected shape of a cantilever beam under bending causing tension at top.

The first step in the problem is to write out the differential equations used when in the double integration method. *EI* is put on the left side of the equation to isolate *M*, which will be integrated.

$$\frac{d^2y}{dx^2} = \frac{Mx}{EI} \qquad EI\frac{d^2y}{dx^2} = Mx$$

M is then replaced by the moment that occurs at any position x in the beam, which in this case is equal to $-PL + Px$, where x is the distance the point is away from the origin, A.

$$EI\frac{d^2y}{dx^2} = -PL + Px$$

The next step is to integrate the equation twice. The first integration gives the following equation:

$$EI\frac{dy}{dx} = -PLx + \frac{Px^2}{2} + C_1 \tag{2.1}$$

The second integration gives this equation.

$$EIy = \frac{-PLx^2}{2} + \frac{Px^3}{6} + C_1x + C_2 \tag{2.2}$$

Now the only thing left to do is to find C_1 and C_2. However, to do this, more information is needed. This information is obtained from the original figure by looking at the boundary conditions of the beam, which in this case show that the left side is fixed and the right side is free.

Because the left end of the beam is fixed, there is no vertical deflection at this point. Therefore, the following equation can be written (where x is the distance from point A and y is the deflection at a point on the beam):

$$When\ x = 0,\ y = 0$$

It is now possible to substitute these values into equation (2.2) and solve for C_2.

$$EI(0) = \frac{-PL(0)^2}{2} + \frac{P(0)^3}{6} + C_1(0) + C_2$$
$$C_2 = 0$$

To solve for C_1, a similar logic is used. Since at the fixed end there is no slope of the beam either, the following equation can be written (where x is the distance from point A and y is the slope at a point on the beam):

$$When\ x = 0,\ \frac{dy}{dx} = 0$$

It is now possible to substitute these values into equation (2.1) and solve for C_1.

$$EI(0) = -PL(0) + \frac{P(0)^2}{2} + C_1$$
$$C_1 = 0$$

Since both C_1 and C_2 are 0, a general expression for deflection of the cantilever beam can now be written from equation (2.2).

$$EIy = \frac{-PLx^2}{2} + \frac{Px^3}{6} + (0)x + (0)$$
$$EIy = \frac{-PLx^2}{2} + \frac{Px^3}{6}$$
$$y(x) = \left(\frac{1}{EI}\right)\left(\frac{-PLx^2}{2} + \frac{Px^3}{6}\right)$$

To find the deflection at the free end B, all that is needed is let $x=L$ in the general expression, since that is the location of the free end.

$$y(L) = \left(\frac{1}{EI}\right)\left(\frac{-PL(L)^2}{2} + \frac{P(L)^3}{6}\right)$$
$$y = \frac{-PL^3}{3EI}$$

Since y is chosen to be upward, the negative sign on the answer indicates that the deflection is downward at point B.

EXAMPLE 2.1.2

Find the expression for the deflection of a simply supported beam subjected to a uniformly distributed load as shown in Figure 2.3. Use the double integration method. The flexural rigidity EI is constant. Choose the pinned support as the origin.

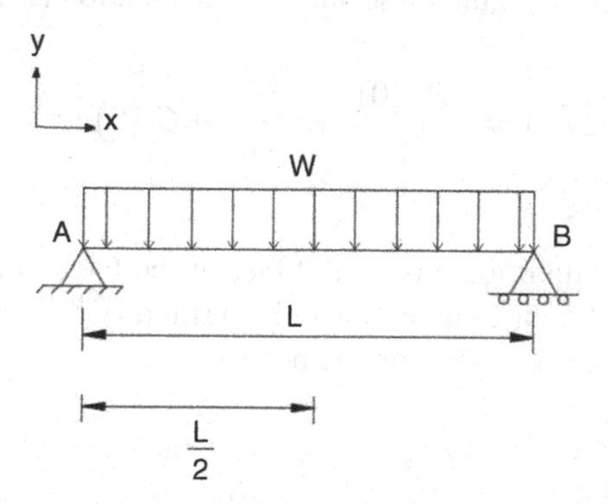

FIGURE 2.3 Simply Supported Beam under Uniform Loading.

Similar to the previous example, the first step in the problem is to write out the differential equation used when in the double integration method. EI is put on the left side of the equation to isolate M, which will be integrated.

$$\frac{d^2y}{dx^2} = \frac{Mx}{EI} \qquad EI\frac{d^2y}{dx^2} = Mx$$

The next step is to again replace Mx with an equation that represents the moment at any point in the beam. This is done by solving for the reaction at A and then finding the resulting moment from the reaction as well as the distributed load. In this example, it produces the following equation:

$$EI\frac{d^2y}{dx^2} = \frac{wL}{2}x - \frac{wx^2}{2}$$

The next step is to integrate the equation twice. The first integration gives the following equation:

$$EI\frac{dy}{dx} = \frac{wL}{2}\frac{x^2}{2} - \frac{wx^3}{6} + C_1 \qquad (2.3)$$

The second integration gives the following equation.

$$EIy = \frac{wL}{4}\frac{x^3}{3} - \frac{w}{6}\frac{x^4}{4} + C_1x + C_2$$
$$EIy = \frac{wLx^3}{12} - \frac{wx^4}{24} + C_1x + C_2 \qquad (2.4)$$

The next step is to again solve to C_1 and C_2 using boundary conditions. Although the left end is not fixed as in the last example, there is no vertical deflection at pinned joints as well. Therefore, the following equation can be written:

$$When\ x = 0,\ y = 0$$

where x is the distance from point A and y is the deflection at a point on the beam. It is now possible to substitute these values into equation (2.4) and solve for C_2.

$$EI(0) = \frac{wL(0)^3}{12} - \frac{w(0)^4}{24} + C_1(0) + C_2$$
$$C_2 = 0$$

Solving for C_1 must be done in a way different from example 2.1.1 as dx/dy cannot be said to equal zero at the location of a pinned joint. However, the right end also has a roller support, which also prevents any vertical deflection at that point. Therefore, the following equation can be written:

$$When\ x = L,\ y = 0$$

where x is the distance from point A and y is the deflection at a point on the beam.

It is now possible to substitute these values into equation (2.4) and solve for C_1.

$$EI(0) = \frac{wL(L)^3}{12} - \frac{w(L)^4}{24} + C_1(L) + (0)$$
$$0 = \frac{wL^4}{12} - \frac{wL^4}{24} + C_1(L)$$
$$-C_1(L) = \frac{wL^4}{24}$$
$$C_1 = -\frac{wL^3}{24}$$

Since both C_1 and C_2 are now found, they can be substituted into equation (2.4) to create a general equation for the entire beam.

$$EIy = \frac{wLx^3}{12} - \frac{wx^4}{24} - \frac{wL^3}{24}x$$
$$y(x) = \left(\frac{w}{EI}\right)\left(\frac{Lx^3}{12} - \frac{x^4}{24} - \frac{L^3}{24}x\right)$$

Example 2.1.3

Find the expression for the deflection of a simply supported beam subjected to a triangular load shown in Figure 2.4. Use the double integration method.

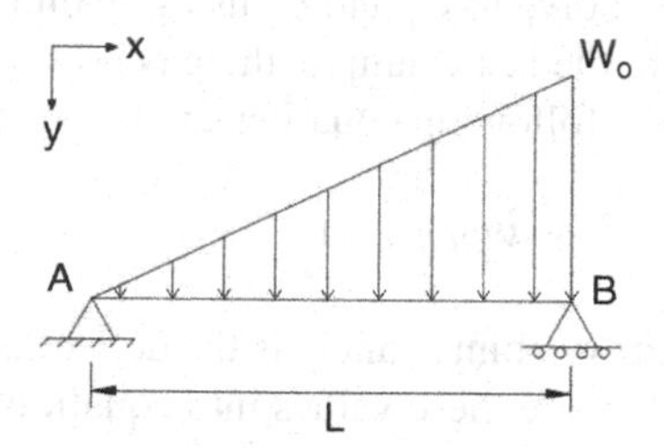

FIGURE 2.4 Simply supported beam under triangular load.

The double integration method can be expanded by starting the integration on the actual loading applied to the beam, instead of starting with the moments at a point. This method requires more than just two integrations and requires the solving of more unknown constants, but will obtain the same result as before.

The first step in this problem is to write out an equation that represents the amount of force at a point on the beam, which in this problem is equal to $w_0 \frac{x}{L}$. Unlike the previous double integration problems, this cannot be set equal to $EI\frac{d^2y}{dx^2}$ then being solved. The reason for this is that $w_0 \frac{x}{L}$ is an equation for the loading at a point on the beam, not the moment at a certain point on the beam. However, these two values are related. When the loading equation is integrated twice, the result is the bending moment equation. Therefore, $w_0 \frac{x}{L}$ must be set equal to $EI\frac{d^4y}{dx^4}$ to give $EI\frac{d^2y}{dx^2}$ when solving for the moment equation.

$$EIy^{IV} = w_0 \frac{x}{L} \tag{2.5}$$

The next step is to integrate equation (2.5) four times in order to obtain an expression for deflection.

$$EIy^{III} = w_0 \frac{x^2}{2L} + C_1 \tag{2.6}$$

$$EIy^{II} = w_0 \frac{x^3}{6L} + C_1 x + C_2 \tag{2.7}$$

$$EIy^{I} = w_0 \frac{x^4}{24L} + C_1 \frac{x^2}{2} + C_2 x + C_3 \tag{2.8}$$

$$EIy = w_0 \frac{x^5}{120L} + C_1 \frac{x^3}{6} + C_2 \frac{x^2}{2} + C_3 x + C_4 \tag{2.9}$$

The last step in the problem is to find C_1, C_2, C_3, and C_4 using the boundary conditions. Similar to the previous example, there can be no vertical deflection at the location of either of the two joints. Therefore,

$$When\, x = 0, x = L \quad y = 0$$

This boundary condition can be used to solve for C_4 letting x=0 when y=0 in equation (2.9).

$$EI(0) = w_0 \frac{(0)^5}{120L} + C_1 \frac{(0)^3}{6} + C_2 \frac{(0)^2}{2} + C_3(0) + C_4$$
$$C_4 = 0$$

Unfortunately, these boundary conditions cannot be used to solve for any other constants as there are three other unknowns in this problem (C_1, C_3, and C_4).

More boundary conditions can be found by knowing this, along with there being no deflection at the two joints, and there can also be no moment at these points. Therefore,

$$When\, x = 0, x = L \quad M = 0$$

This can be used in equation (2.7), as y^{II} is equivalent to M. Therefore, x=0 when M=0 can be substituted in to find the following:

$$EI(0) = w_0 \frac{(0)^3}{6L} + C_1(0) + C_2$$
$$C_2 = 0$$

Now that C_2 is found, $x=L$ when $M=0$ can be substituted into equation (2.7) to solve for C_1.

$$EI(0) = w_0 \frac{(L)^3}{6L} + C_1(L) + 0$$
$$0 = w_0 \frac{L^2}{6} + C_1(L)$$
$$C_1 = -\frac{w_0 L}{6}$$

C_3 can now be found by letting $x=L$ when $y=0$ in equation (2.9).

$$EI(0) = w_0 \frac{(L)^5}{120L} + \left(-\frac{w_0 L}{6}\right)\frac{(L)^3}{6} + (0)\frac{(L)^2}{2} + C_3(L) + (0)$$
$$0 = w_0 \frac{L^4}{120} + \left(-\frac{w_0 L}{6}\right)\frac{(L)^3}{6} + C_3(L)$$
$$C_3 = \frac{7L^3 w_0}{360}$$

Finally, equation (2.9) can be rewritten with the constants replaced with the correct values.

$$y(x) = \frac{1}{EI}\left(\frac{w_0 x^5}{120L} - \frac{w_0 L x^3}{6} + \frac{7w_0 L^3 x}{360}\right)$$

EXAMPLE 2.1.4

Derive the general expression for deflection of a simply supported beam subjected to concentrated load as shown in Figure 2.5. Use the double integration method.

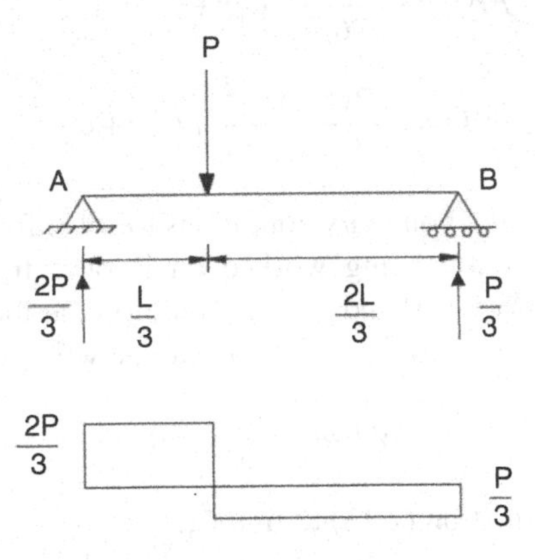

FIGURE 2.5 Reactions and shear of a simply supported Beam under concentrated load at distance L/3 from A.

This problem requires a slightly different approach than the previous three examples because the loading on the beam is neither uniformly distributed nor located on either end. This means that it is impossible to write a single equation that works for the entire span of the beam. Things such as loads, joints, and supports cause this discontinuity in the beam and prevent it from being just one equation. The solution to this problem is to simply split the problem into two by writing one equation before the load is applied (i.e., from $0<x<L/3$) and one equation after the load is applied ($L/3<x<L$).

Since the shear force diagram is given in this problem, a method similar to example 2.1.3 will be used to start the integration. The first section that will be done in this problem is the span between $0<x<L/3$. Since y^{III} correlates to the shear force at a point in the beam, and the shear force is constant for this part of the beam, the following equation can be written:

$$\begin{gathered} From\, 0 < x < \frac{L}{3}, \\ EIy^{III} = -V = -\frac{2P}{3} \end{gathered} \tag{2.10}$$

The next step is to integrate this equation three times in order to get a general expression of deflection.

$$From\, 0 < x < \frac{L}{3},$$

$$EIy^{II} = -\frac{2Px}{3} + C_1 \tag{2.11}$$

$$EIy^{I} = -\frac{2Px^2}{6} + C_1x + C_2 \tag{2.12}$$

$$EIy = -\frac{Px^3}{9} + \frac{C_1x^2}{2} + C_2x + C_3 \tag{2.13}$$

The next step is to use the boundary conditions to solve for the constants. Because the segment that is currently being worked on is only from $0<x<L/3$, boundary conditions are applied when $x=0$ and $x=L/3$. Therefore, at the pinned support ($x=0$), there is no vertical deflection, so the following can be written:

$$When\, x = 0, y = 0$$

This can be used in equation (2.13) to find C_3.

$$EI(0) = -\frac{P(0)^3}{9} + \frac{C_1(0)^2}{2} + C_2(0) + C_3$$
$$C_3 = 0$$

Since there are also no moments at the pinned support either, the following can also be written:

$$When\, x = 0, M = 0$$

Then, by substituting these values into equation (2.11), C_1 can be found.

$$EI(0) = -\frac{2P(0)}{3} + C_1$$
$$C_1 = 0$$

C_2 cannot be found without first solving part of the segment (as there is no boundary condition that relates C_2 to y^I), so the next step is to begin the segment of $L/3<x<L$. Similar to the first segment, the shear force is a constant and the following equation can be written:

$$From\, \frac{L}{3} < x < L,$$

$$EIy^{III} = -V = \frac{P}{3} \tag{2.14}$$

Then, this equation is integrated three times in order to get a general expression of deflection.

$$From\ \frac{L}{3} < x < L,$$

$$EIy^{II} = \frac{Px}{3} + C_4 \tag{2.15}$$

$$EIy^{I} = \frac{Px^2}{6} + C_4 x + C_5 \tag{2.16}$$

$$EIy = \frac{Px^3}{18} + \frac{C_4 x^2}{2} + C_5 x + C_6 \tag{2.17}$$

Since the boundary conditions are from $L/3<x<L$ in this segment, and at the roller support ($x=L$) there is no moment or vertical deflection, the following can be written:

$$When\ x = L,\ M = 0$$
$$When\ x = L,\ y = 0$$

This can then be used with equation (2.15) to find C_4.

$$EI(0) = \frac{PL}{3} + C_4$$
$$C_4 = -\frac{PL}{3}$$

Letting $x=L$ when $y=0$ in equation (2.17) does not solve for either C_5 or C_6, but it does give an equation relating the two which will be used later in the problem.

$$EI(0) = \frac{P(L)^3}{18} + \left(-\frac{PL}{3}\right)\frac{(L)^2}{2} + C_5(L) + C_6$$

$$PL^3\left(\frac{1}{6} - \frac{1}{18}\right) = C_5 L + C_6 \tag{2.18}$$

$$C_5 L + C_6 = \frac{1}{9}PL^3$$

To find the remaining constants (C_2, C_5, and C_6), a relationship between the two segments must be found. This can be done easily as both share a common point in $L/3$. At this point, the deflection is the same and the slope is the same for both segments. This means that the following equation can be written:

$$At\ x = \frac{L}{3}, y = y\ and\ y^I = y^I$$

This means that it is possible to set equations (2.12) and (2.16) together to obtain an equation relating C_2 and C_5.

$$-\frac{2P\left(\frac{L}{3}\right)^2}{6}+(0)x+C_2=\frac{P\left(\frac{L}{3}\right)^2}{6}+\left(-\frac{PL}{3}\right)\left(\frac{L}{3}\right)+C_5$$

$$-\frac{2PL^2}{54}+C_2=\frac{PL^2}{54}+\left(-\frac{PL}{3}\right)\left(\frac{L}{3}\right)+C_5 \tag{2.19}$$

$$C_2-C_5=-\frac{PL^2}{18}$$

The next step is to set equations (2.13) and (2.17) together to obtain C_6.

$$-\frac{P\left(\frac{L}{3}\right)^3}{9}+\frac{(0)x^2}{2}+C_2x+(0)=\frac{P\left(\frac{L}{3}\right)^3}{18}+\frac{-\frac{PL}{3}\left(\frac{L}{3}\right)^2}{2}+C_5\left(\frac{L}{3}\right)+C_6$$

$$C_2-C_5=\frac{3}{L}\left(\frac{PL^3}{162}+\frac{-PL^3}{54}+C_6\right)$$

Use equation (2.19) to remove C_2 and C_5.

$$-\frac{PL^2}{18}=\frac{3}{L}\left(\frac{PL^3}{162}+\frac{-PL^3}{54}+C_6\right)$$

$$C_6=-\frac{PL^3}{162}$$

Now using equation (2.18), C_5 can be found.

$$C_5L-\frac{PL^3}{162}=\frac{1}{9}PL^3$$

$$C_5=\left(\frac{1}{9}+\frac{1}{162}\right)PL^2$$

$$C_5=\frac{19}{162}PL^2$$

Next, C_2 can be found using equation (2.19).

$$C_2-\frac{19}{162}PL^2=-\frac{PL^2}{18}$$

$$C_2=\frac{5}{81}PL^2$$

The last step is to finally write the complete general expression for deflection for both segments of the beam.

Note: The above problem can be solved without splitting at the concentrated load location by introducing Dirac Delta function, which is beyond the scope of this textbook.

2.2 MOMENT AREA METHOD

There are two theorems that best describe this method.

Theorem I The change in slope between the tangents to the elastic curve at two points in a straight member under bending (Figure 2.6a) is equal to the area of M/EI diagram between those two points. It can be expressed mathematically as,

$$\Delta\theta_{AB} = \int_A^B d\theta = \int_A^B \frac{M}{EI} dx \tag{2.20}$$

Theorem II The deflection of a point on a straight member under bending in the direction perpendicular to the original straight axis of the member, measured from the tangent at another point on the member is equal to the moment of the M/EI diagram between those two points about the point where the deflection occurs (Figure 2.6b). It can be expressed mathematically as,

$$t_{BA} = \int_A^B x\,d\theta = \int_A^B \frac{Mx}{EI} dx \tag{2.21}$$

The following examples illustrate the application of moment-area method to various beams and a couple of frames.

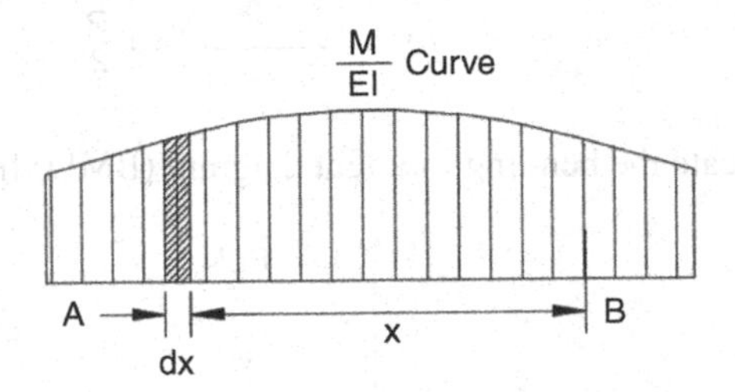

FIGURE 2.6a M/EI curve along the Beam Length.

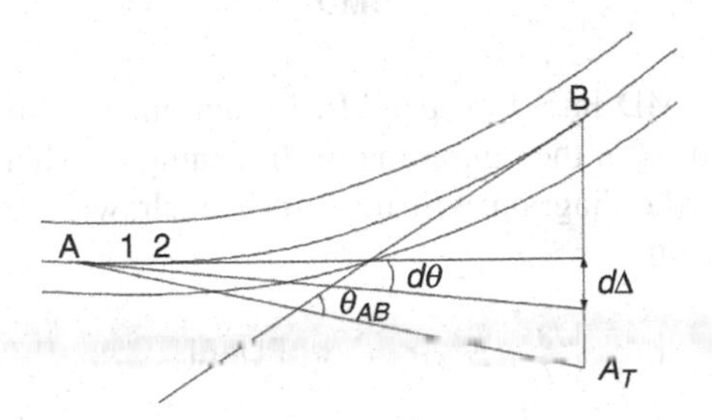

FIGURE 2.6b Deflected Shape and Slopes between parts A & B.

Example 2.2.1

Find the slopes at the point A and B (θ_{AB}) and the deflection at B (δ_B) for the simply supported beam shown in Figure 2.7. Use the moment-area method. Assume EI is constant for the entire beam.

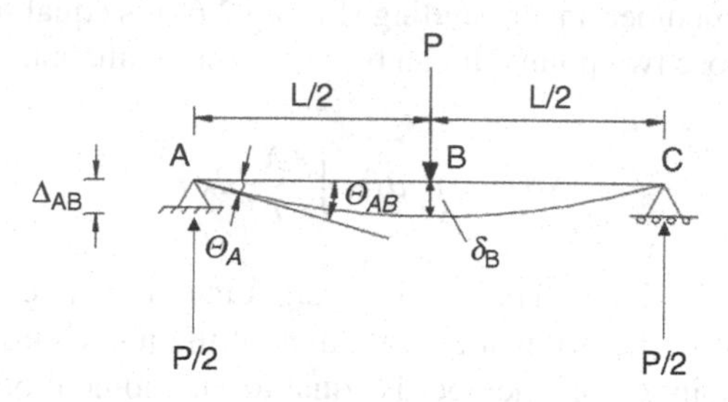

FIGURE 2.7 Deflected shape of Simply Supported Beam with Load at the Center.

The first step in solving the problem is to create the shear force diagram (SFD) for the beam.

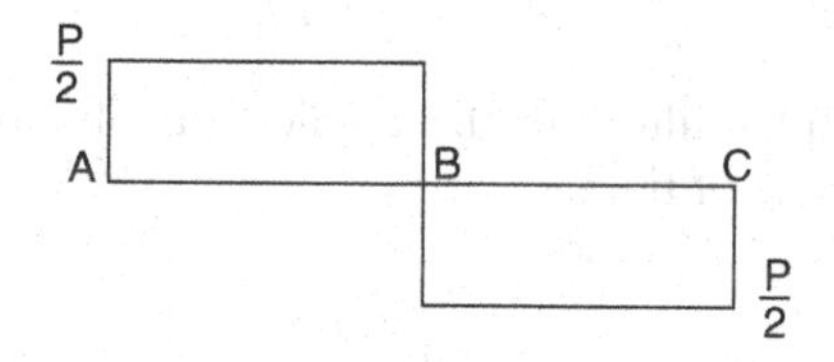

The next step is to create the bending moment diagram (BMD) from the SFD.

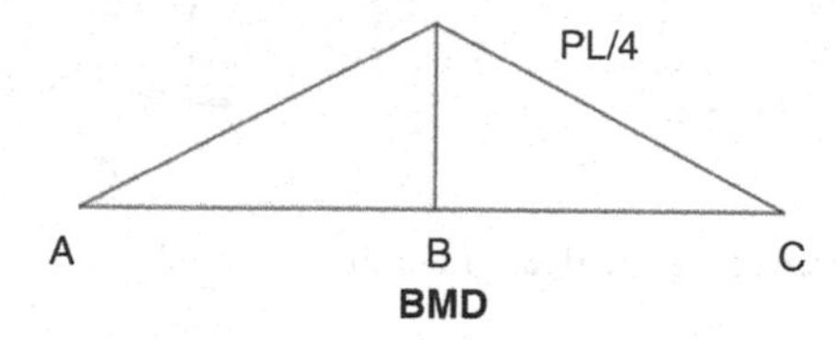

Next, convert the BMD into a proper M/EI diagram. For this example, the only thing needed is to add EI in the denominator. In examples where EI is not constant throughout the beam, the diagram will need to be redrawn to accurately show the differences in the EI values.

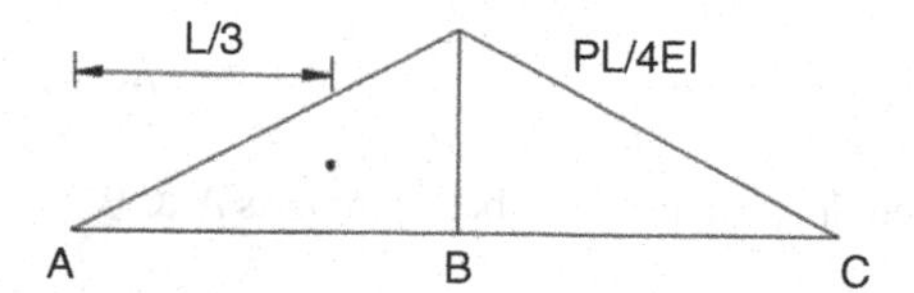

Using Theorem 1, which states that the change in slope in a straight member under bending is equal to the area of M/EI diagram between those two points, the slope at A and B (θ_{AB}) can be found by the following equation

$$\theta_A = \theta_{AB} = \frac{1}{2}\left(\frac{L}{2}\right)\left(\frac{PL}{4EI}\right) = \frac{PL^2}{16EI}$$

The deflection at B (δ_B) is found by implementing Theorem 2, which states that the deflection on a point on the member is equal to the moment of the M/EI diagram at that point. Therefore, the following equation can be written:

$$\delta_B = \Delta_{AB} = \frac{1}{2}\left(\frac{L}{2}\right)\left(\frac{PL}{4EI}\right)\left(\frac{L}{3}\right) = \frac{PL^3}{48EI}$$

Example 2.2.2

Find the slope at A of a simply supported beam subjected to a concentrated load and a moment at support B as shown in Figure 2.8. Use the moment-area method.

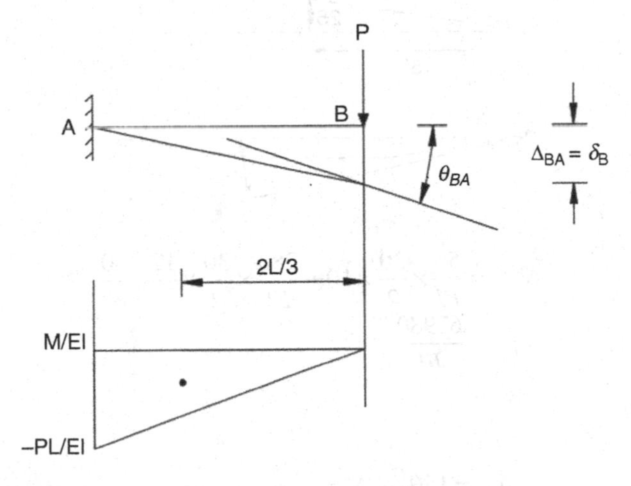

FIGURE 2.8 Slope & moment of a simply supported beam.

Figure 2.8 shows the

$$\theta_{BA} = \frac{1}{2}(L)\left(-\frac{PL}{EI}\right) = -\frac{PL^2}{2EI}$$

$$\Delta_{BA} = \delta_B = \frac{1}{2}(L)\left(-\frac{PL}{EI}\right)\left(\frac{2L}{3}\right) = -\frac{PL^3}{3EI}$$

Example 2.2.3

Find the maximum deflection for a simply supported beam as shown in Figure 2.9. Use the moment-area method.

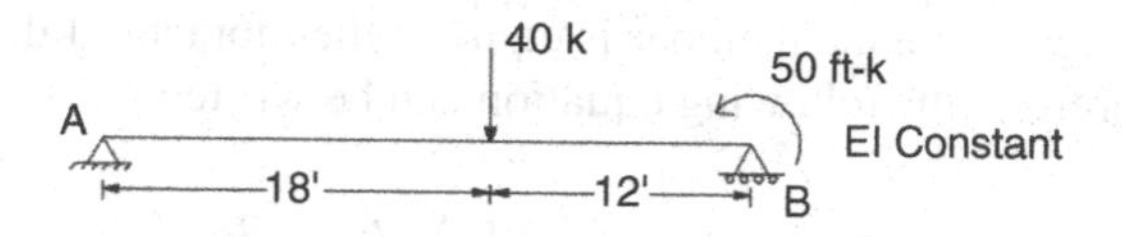

FIGURE 2.9 Simply supported (S.S.) beam under moment & concentrated load.

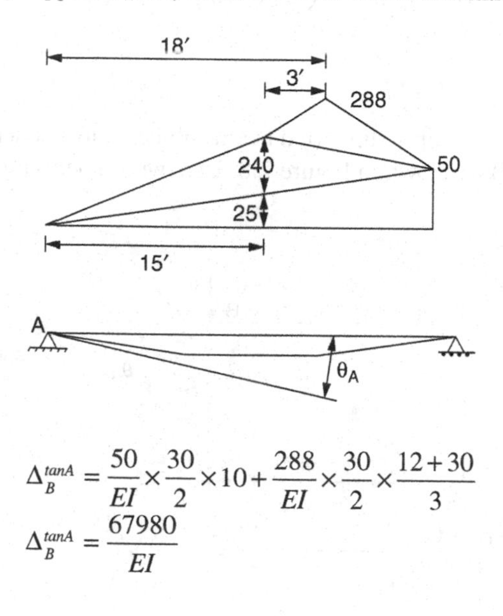

$$\Delta_B^{tanA} = \frac{50}{EI} \times \frac{30}{2} \times 10 + \frac{288}{EI} \times \frac{30}{2} \times \frac{12+30}{3}$$

$$\Delta_B^{tanA} = \frac{67980}{EI}$$

Compute δ_{CL}

$$\delta_{\mathrm{CL}} = 15\theta_A - \Delta_{\mathrm{CL}}^{tanA}$$

$$\delta_{\mathrm{CL}} = \frac{15(2266)}{EI} - 265 \times \frac{15}{2} \times 5 \times \frac{1}{EI}$$

$$\delta_{\mathrm{CL}} = \frac{24052}{EI} \downarrow$$

Example 2.2.4

Find the maximum deflection for a simply supported beam as shown in Figure 2.10. Use the moment-area method.

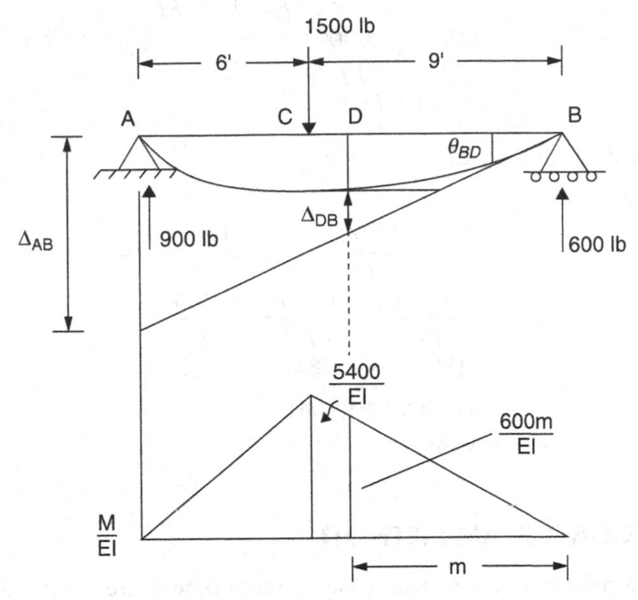

FIGURE 2.10 S.S. beam with concentrated load.

$$\frac{5400}{9EI} = \frac{y}{m}$$

$$y = \frac{m}{9}\left(\frac{5400}{EI}\right) = \frac{600m}{EI}$$

$$\Delta_{DB} = \frac{1}{2}\left(\frac{600m}{EI}\right)(m)\left(\frac{m}{3}\right) = \frac{600m^3}{6EI} = \frac{100m^3}{EI}$$

$$\Delta_{AB} = \frac{1}{2}\left(\frac{5400}{EI}\right)(6)(4) + \frac{1}{2}\left(\frac{5400}{EI}\right)(9)(9)$$

$$\Delta_{AB} = \frac{64,800}{EI} + \frac{218,700}{EI} \tag{2.22}$$

$$\Delta_{AB} = \frac{283,500}{EI}$$

$$\theta_{BD} = \frac{\Delta_{AB}}{15} = \frac{283,500}{15EI} = \frac{18,900}{EI}$$

$$\delta_D + \Delta_{BD} = m(\theta_{BD}) = m\frac{18,900}{EI}$$

$$\delta_D = \frac{18,900}{EI}m - \frac{100m^3}{EI}$$

Need expression to solve for *m*.

$$\theta_{BD} \text{ is also} = \frac{1}{2}(m)\left(\frac{600m}{EI}\right) = \frac{300m^2}{EI}$$

$$\frac{300m^2}{EI} = \frac{18,900}{EI}$$

$$m = 7.93'$$

Using the above equation (2.22)

$$\delta_D = \delta_{max} = \frac{18,900}{EI}(7.93) - \frac{100(7.93)^3}{EI}$$

$$\delta_{max} = \frac{150,000}{EI} - \frac{50,200}{EI} = \frac{99,800}{EI}$$

$$\delta_{max} = \frac{(99,800)(1728)}{(1,760,000)(108)}$$

$$\delta_{max} = 0.907''$$

2.3 CONJUGATE BEAM METHOD

This can be considered as a special case of the moment-area method. For some problems, it might be easier to use conjugate beam method than moment-area method.

In the moment-area method, it is important that the beam considered has a point where there is a horizontal tangent. If such a point does not exist, then it is easier to use conjugate beam method. In the conjugate beam method, for every real beam there exists a conjugate beam. There is what is called a support equivalence. A fixed support in a real beam becomes a free support in a conjugate beam, while a simple support remains the same. These equivalent supports are shown in Figure 2.11c. This equivalence of supports is inherently based on the following two equations (see Figures 2.11a and 2.11b).

$$\theta_C = V'_C \tag{2.23}$$

$$\Delta_C = M'_C \tag{2.24}$$

The following examples illustrate the use of conjugate beam method to various beams.

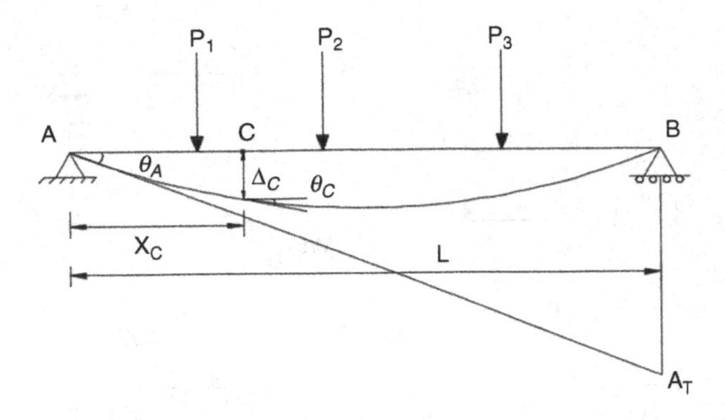

FIGURE 2.11a S.S. beam under concentrated loads.

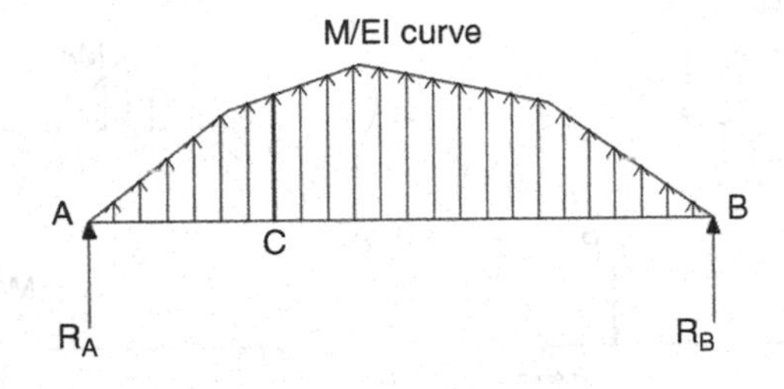

FIGURE 2.11b Conjugate beam of Figure 2.11a.

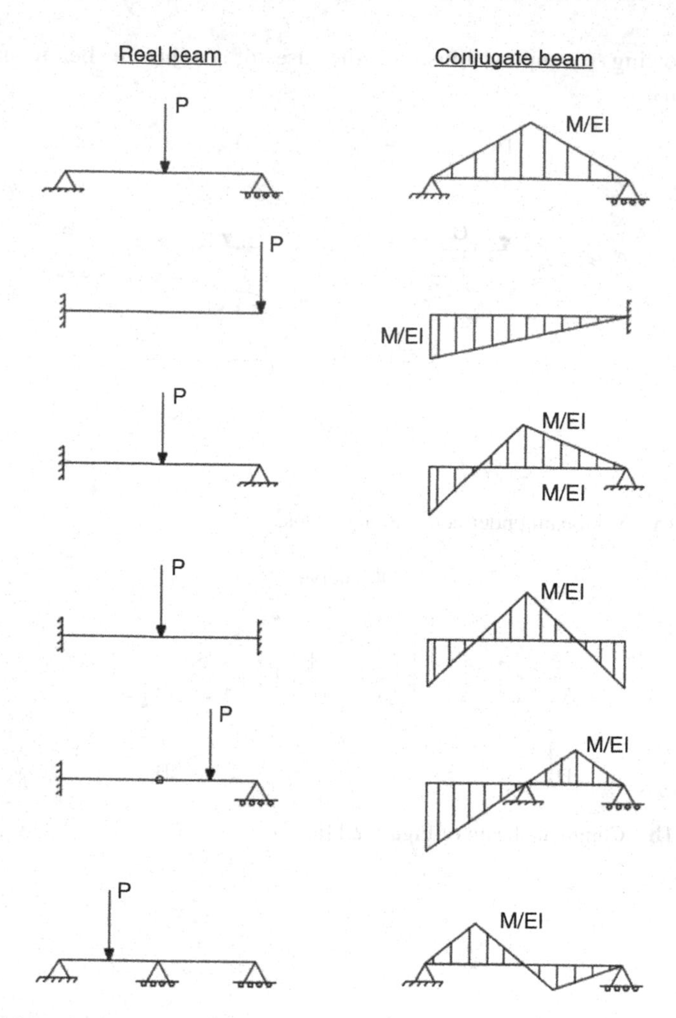

FIGURE 2.11C Support equivalence of Real Vs Conjugate Beam.

Example 2.3.1

Find all the reactions for the beam (with variable moment of inertia), acted on by a uniformly distributed load as well as a concentrated load, as shown in Figure 2.12. Use the conjugate beam method. EI=2.4×10[7] in[2]-k.

Also, draw the final SFD and BMD.

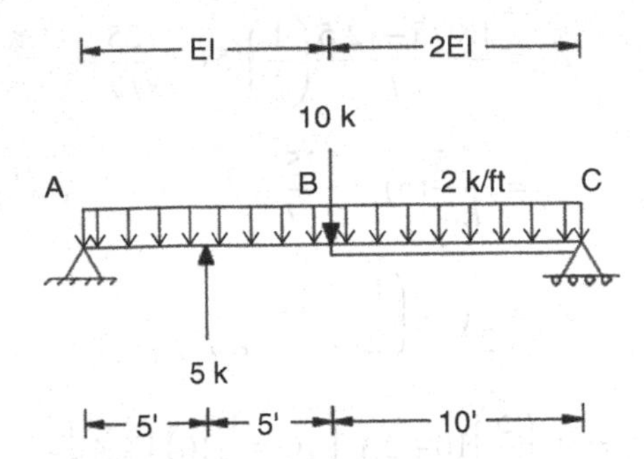

FIGURE 2.12 S.S. beam of varying M.I. under different load types.

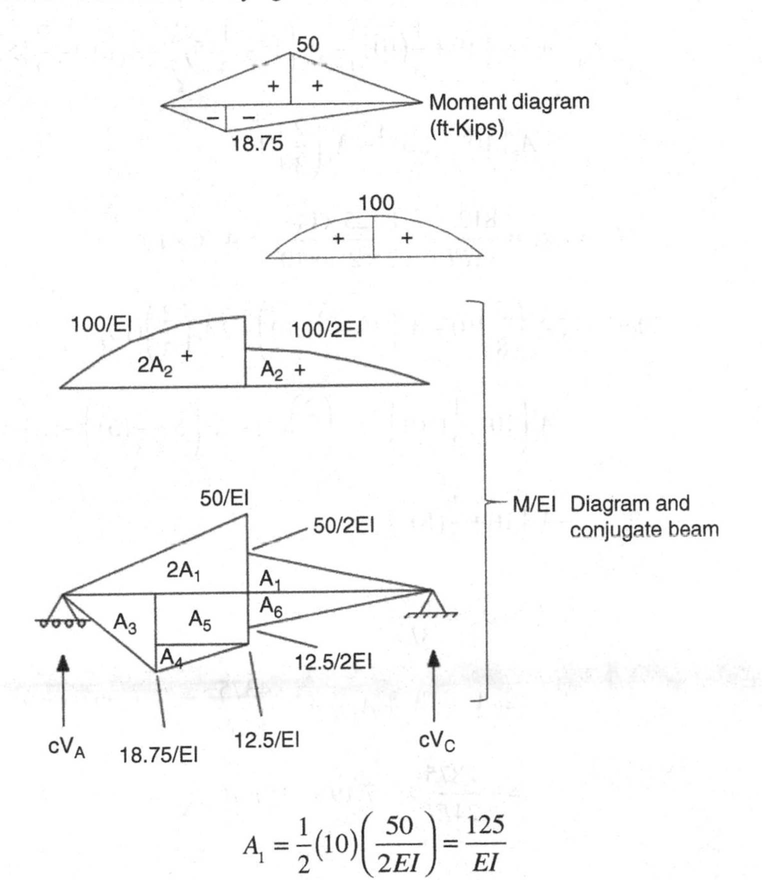

$$A_1 = \frac{1}{2}(10)\left(\frac{50}{2EI}\right) = \frac{125}{EI}$$

$$A_2 = \frac{2}{3}(10)\left(\frac{100}{2EI}\right) = \frac{1000}{3EI}$$

$$A_3 = \frac{1}{2}(5)\left(\frac{18.75}{EI}\right) = \frac{375}{8EI}$$

$$A_4 = \frac{18.75-12.5}{EI}\left(\frac{1}{2}\right)(5) = \frac{125}{8EI}$$

$$A_5 = \frac{12.5}{EI}(5) = \frac{125}{2EI}$$

$$A_6 = \frac{1}{2}(10)\left(\frac{12.5}{2EI}\right) = \frac{125}{4EI}$$

$$20cV_A = A_2\left(\frac{5}{8}\right)10 + 2A_2\left(10+\frac{3}{8}(10)\right) + A_1\left(\frac{2}{3}\right)10$$

$$+2A_1\left(10+\frac{1}{3}(10)\right) - A_3\left(15+\frac{1}{3}(5)\right) - A_4\left(10+\frac{2}{3}(5)\right)$$

$$-A_5\left(10+\frac{1}{2}(5)\right) - A_4\left(\frac{2}{3}\right)(10)$$

$$\theta_A = cV_A = \frac{8125}{12EI} = \frac{8125\times 144}{12\times 2.4\times 10^7} = 4.06\times 10^{-3}\,rad$$

$$20cV_C = 2A_2\left(\frac{5}{8}\right)10 + A_2\left(10+\frac{3}{8}(10)\right) + 2A_1\left(\frac{2}{3}\right)(10)$$

$$+A_1\left(10+\frac{1}{3}(10)\right) - A_3\left(\frac{2}{3}\right)(5) - A_4\left(5+\frac{1}{3}(5)\right) - A_5\left(5+\frac{5}{2}\right)$$

$$-A_6\left(10+\frac{1}{3}(10)\right)$$

$$cV_C = \theta_C = \frac{4375}{8EI}$$

$$\theta_B = cV_B = A_1 + A_2 - A_6 - \frac{4375}{8EI}$$

$$= -\frac{2875}{24EI} = -7.19\times 10^{-4}\,rad$$

$$cM_B = \frac{4375}{8EI}(10) + A_6\left(\frac{1}{3}\right)(10) - A_1\left(\frac{1}{3}\right)(10) - A_2\left(\frac{3}{8}\right)(10)$$

$$\Delta_B = cM_B = \frac{15625}{4EI} = 2.34 \times 10^{-2}\ ft$$

Example 2.3.2

Find all reactions for the simply supported beam, with constant EI, and acted on by a concentrated load, as shown in Figure 2.13. Draw also SFD and BMD diagrams. Use the conjugate beam method.

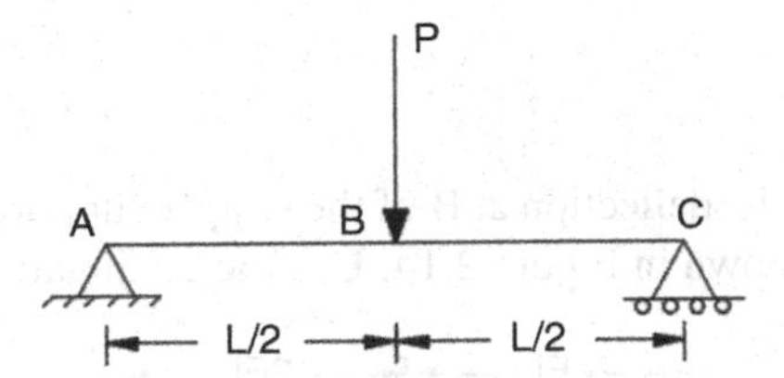

FIGURE 2.13 S.S. beam with concentrated load at mid span.

$$\Sigma M_C = 0 = 2\left[\frac{1}{2}\left(\frac{PL}{4EI}\right)\left(\frac{L}{2}\right)\right]\left(\frac{L}{2}\right) - cV_C L$$

$$cV_C = \frac{PL^2}{16EI}$$

$$\Sigma F_y = 0 = cV_a + cV_C - \frac{PL^2}{8EI}$$

$$\Rightarrow$$

$$cV_a = \frac{PL^2}{16EI} = \theta_A \qquad \textit{in the real beam}$$

PL/4EI

cM_B

PL²/16EI   cV_B

$$cM_B = \frac{PL^3}{32EI} - \left[\frac{PL}{4EI}\left(\frac{1}{2}\right)\left(\frac{L}{2}\right)\right]\left(\frac{1}{3}\left(\frac{L}{2}\right)\right)$$

$$cM_B = \frac{PL^3}{48EI} = \Delta_B \qquad \textit{in the real beam}$$

Example 2.3.3

Find the slope at A and the deflection at B of the simply supported beam with variable moment of inertia, as shown in Figure 2.14. Use the conjugate beam method.

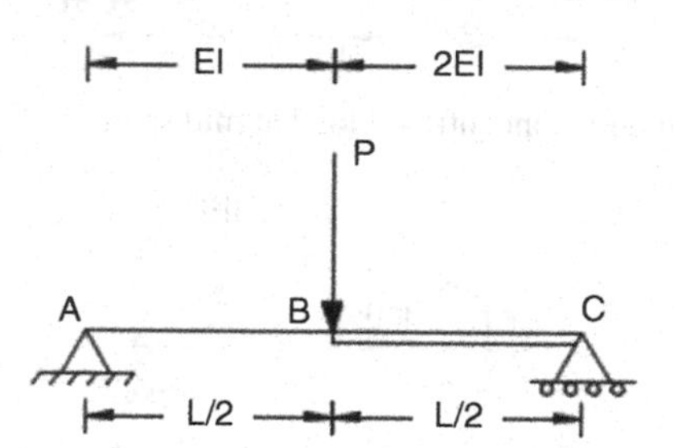

FIGURE 2.14 S.S. beam of varying M.I. under concentrated load at mid span.

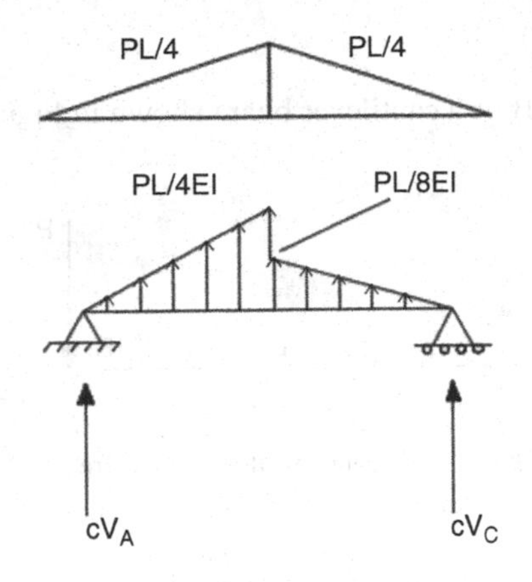

$$\sum M_C = 0 = \left[\frac{1}{2}\left(\frac{PL}{8EI}\right)\left(\frac{L}{2}\right)\right]\left[\left(\frac{2}{3}\right)\left(\frac{L}{2}\right)\right] + \left[\frac{1}{2}\left(\frac{PL}{4EI}\right)\left(\frac{L}{2}\right)\right]\left[\frac{L}{2} + \frac{1}{3}\left(\frac{L}{2}\right)\right] - cV_A L$$

$$cV_A = \frac{5PL^2}{96EI} = \theta_A$$

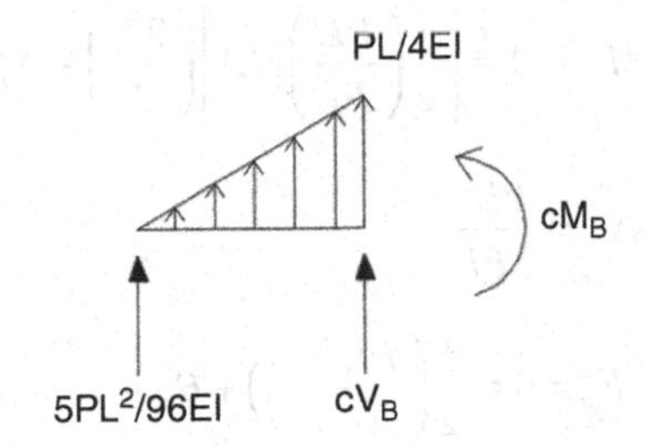

$$cM_B = \Delta_B = \frac{5PL^2}{96EI}\left(\frac{L}{2}\right) - \frac{1}{2}\left(\frac{L}{2}\right)\left(\frac{PL}{4EI}\right)\left(\frac{1}{3}\right)\left(\frac{L}{2}\right)$$

$$\Delta_B = \frac{3PL^3}{192EI}$$

EXAMPLE 2.3.4

Find the slope at A and B of a cantilever beam shown in Figure 2.15. Use the conjugate beam method.

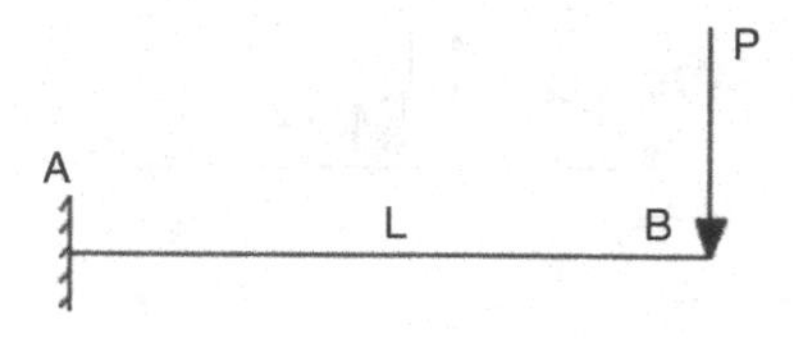

FIGURE 2.15 Cantilever beam with concentrated load at free end.

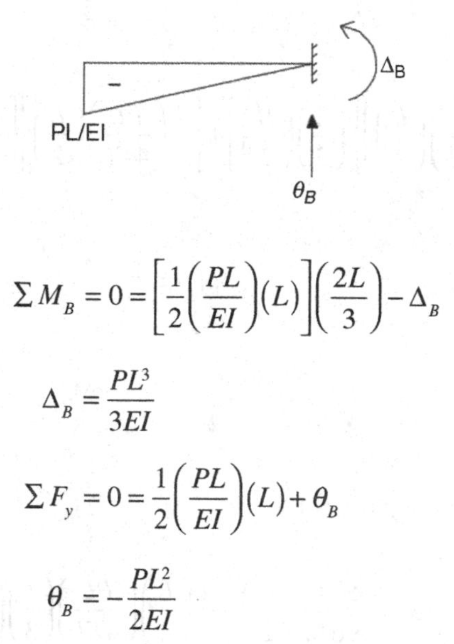

$$\Sigma M_B = 0 = \left[\frac{1}{2}\left(\frac{PL}{EI}\right)(L)\right]\left(\frac{2L}{3}\right) - \Delta_B$$

$$\Delta_B = \frac{PL^3}{3EI}$$

$$\Sigma F_y = 0 = \frac{1}{2}\left(\frac{PL}{EI}\right)(L) + \theta_B$$

$$\theta_B = -\frac{PL^2}{2EI}$$

2.4 UNIT LOAD METHOD

The unit load method is one of the most powerful methods that can be used for finding deflections for statically determinate structures as well as analyzing indeterminate structures by force method (which will be discussed later in Chapter 4 in detail) wherein compatibility of displacements at supports has to be established for the basic determinate structure.

The basic equation for finding slope or deflection at any point of a continuous structure (beam or a frame) is given as follows:

$$\theta \ or \ \Delta = \int_A^B \frac{Mm}{EI} dx \tag{2.25}$$

where M = Bending moment at any section due to applied moment
m = Bending moment at the same section due to unit load

Unit load is applied at the point where the deflection is required in the direction of deflection. This implies that if a vertical deflection is required, a unit vertical load is applied. Similarly, if a horizontal deflection is required, a horizontal unit load is applied. If the final deflection comes out to be positive, that means the assumed direction of deflection is correct.

For a discrete structure (such as a truss), the expression for calculation of deflection at any point in the truss is given as,

$$\Delta = \Sigma \frac{NnL}{AE} \tag{2.26}$$

where N = Forces in the truss due to applied load
n = Forces in the truss due to unit load

Here the term "load" is used in a general sense. It can be a force or a moment also. The most important thing to be noted in the unit load method is that when obtaining "m," one needs to remove all the applied loads and apply a unit force (if deflection at any point in the structure is to be obtained) or a unit moment (if slope at any point in the structure is to be obtained).

A few examples are solved below for finding deflection of determinate beams and frames. The sign convention used is shown in Figure 2.16.

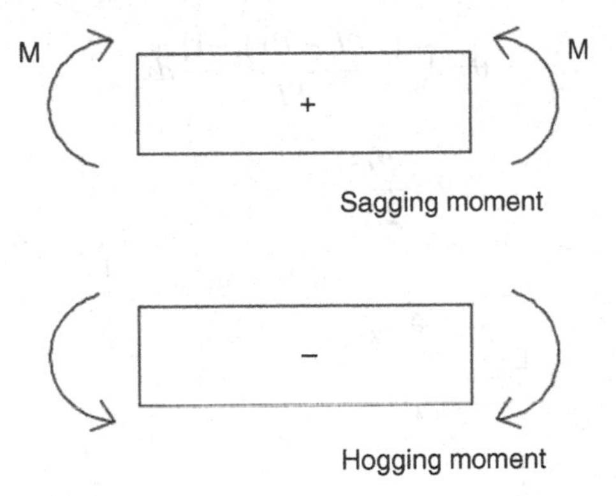

FIGURE 2.16 Sign convention for bending.

Example 2.4.1

Find the slope and deflection of determinate beams and frames. The sign convention used is shown in Figure 2.4. Use the Unit load method.

$$\theta = \int_d^L \frac{Mm}{EI} dx \tag{2.27}$$

$$\Delta = \int_d^L \frac{Mm}{EI} dx \tag{2.28}$$

Slope Calculation Table for θ_B

Portion of Beam	AB
Origin	A
Limits	$x = 0 \; to \; x = L$
M	$-PL + Px$
M	-1

$$\theta = \int_0^L \frac{(-PL + Px)(-1)}{EI} dx$$

$$\theta = \frac{PL^2}{2EI}$$

Deflection calculation for Δ_B

Portion of Beam	AB
Origin	A
Limits	$x = 0 \, to \, x = L$
M	$-PL + Px$
M	$-L + x$

$$\Delta_B = \int_0^L \frac{(-PL + Px)(-L + x)}{EI} dx$$

$$\Delta_B = \frac{PL^3}{3EI} \quad \downarrow$$

EXAMPLE 2.4.2

Find the slope at end A (θ_A) and at B (θ_B) and the deflection at C (Δ_C) of the simply supported beam shown in Figure 2.17. Use the unit load method.

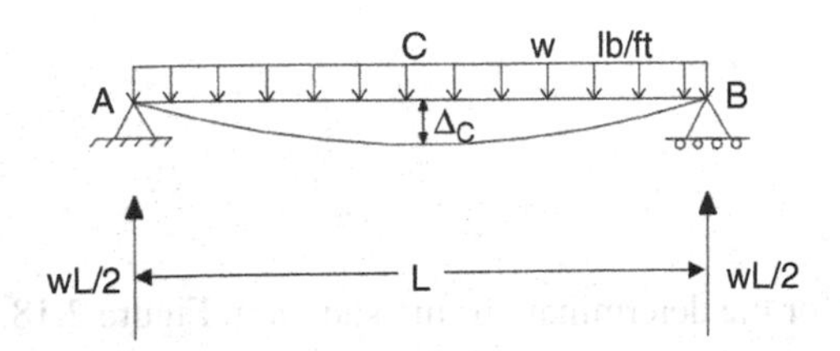

FIGURE 2.17 S.S. beam under uniform load.

Portion of Beam	AB	CB
Origin	A	B
Limits	$x = 0 to x = \frac{L}{2}$	$x = 0 to x = \frac{L}{2}$
M	$\frac{WL}{2}x - \frac{Wx^2}{2}$	$\frac{WL}{2}x - \frac{Wx^2}{2}$
m for θ_A	$1 - \frac{x}{L}$	$\frac{x}{L}$
m for Δ_C	$\frac{1}{2}x$	$\frac{1}{2}x$

$$\theta_A = \frac{1}{EI}\left[\int_0^{L/2}\left(\frac{wL}{2}x - \frac{wx^2}{L}\right)\left(1-\frac{x}{L}\right)dx\right] + \left[\int_0^{L/2}\left(\frac{wL}{2}x - \frac{wx^2}{L}\right)\left(\frac{x}{L}\right)dx\right]$$

$$\theta_A = \frac{wL^3}{24EI}$$

$$\theta_B = -\theta_A = -\frac{wL^3}{24EI}$$

$$\Delta_C = \frac{1}{EI}\left[2\int_0^{L/2}\left(\frac{wL}{2}x - \frac{wx^2}{L}\right)\left(\frac{1}{2}x\right)dx\right]$$

$$\Delta_C = \frac{5wL^4}{384EI} \quad \downarrow$$

EXAMPLE 2.4.3

Find the slope at A for the determinate frame shown in Figure 2.18. Use the unit load method.

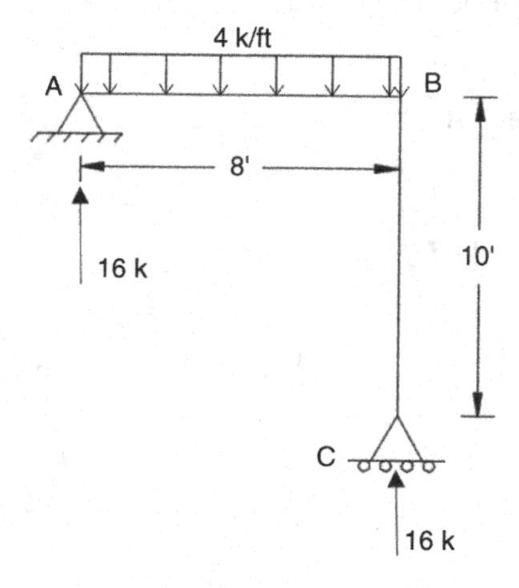

FIGURE 2.18 Beam-column frame with 90 degree bent.

Portion of Beam	AB	BC
Origin	A	C
Limits	$x = 0\ to\ x = 8$	$x = 0\ to\ x = 10$
M	$16x - 2x^2$	0
m for θ_A	$1 - \frac{x}{8}$	0

$$\theta_A = \int_0^8 \frac{Mm}{EI}dx$$

$$\theta_A = \frac{1}{EI}\int_0^8 \left(16x - 2x^2\right)\left(1 - \frac{x}{8}\right)dx$$

$$\theta_A = \frac{256}{3EI}$$

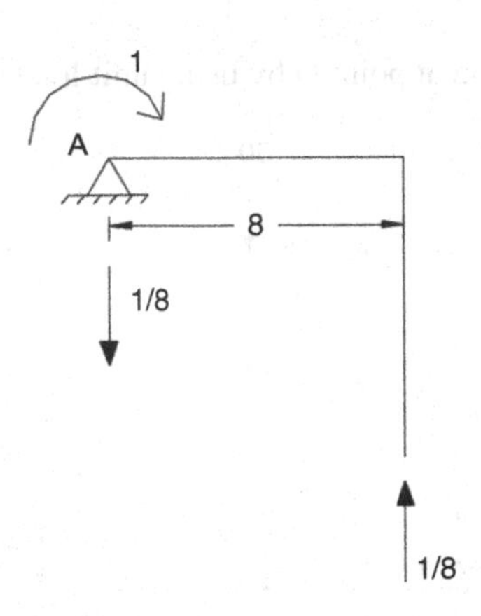

2.5 SUMMARY

This chapter discussed the important methods for calculation of deflection of various determinate structures using four well-known methods: double integration method, moment-area method, conjugate beam method, and the unit load method. Several examples for finding deflection of various structures subjected to different kinds of loading are discussed.

PROBLEMS FOR CHAPTER 2

Problem 2.1 Find deflection at point C by using unit load method.

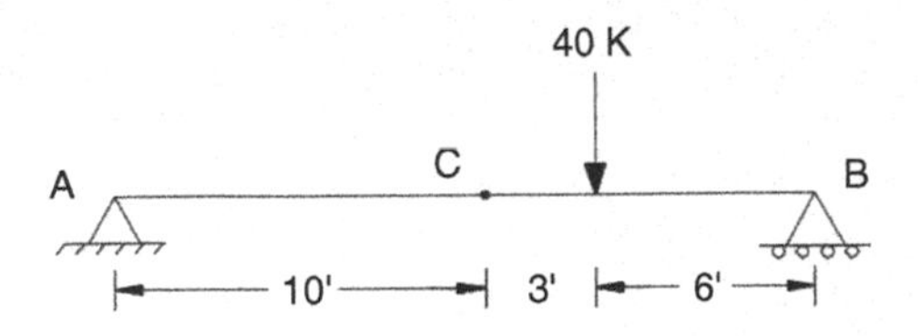

Problem 2.2 Find deflection at point C by using moment-area method.

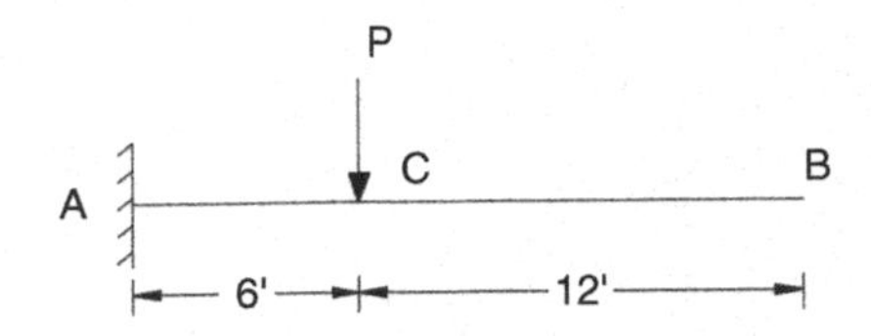

Problem 2.3 Find deflection at point C by using unit load method.

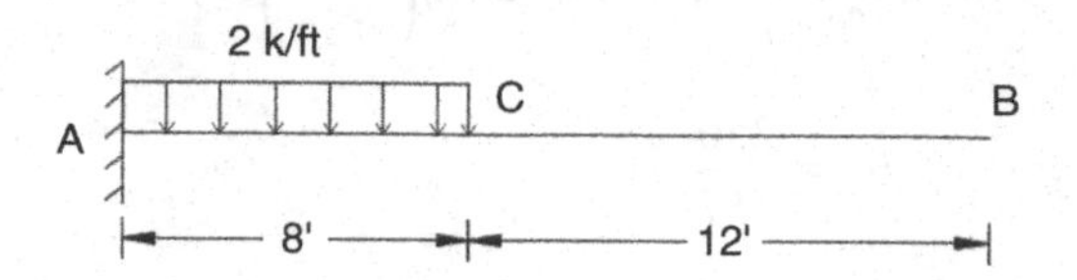

Problem 2.4 Find deflection at point D by using unit load method.

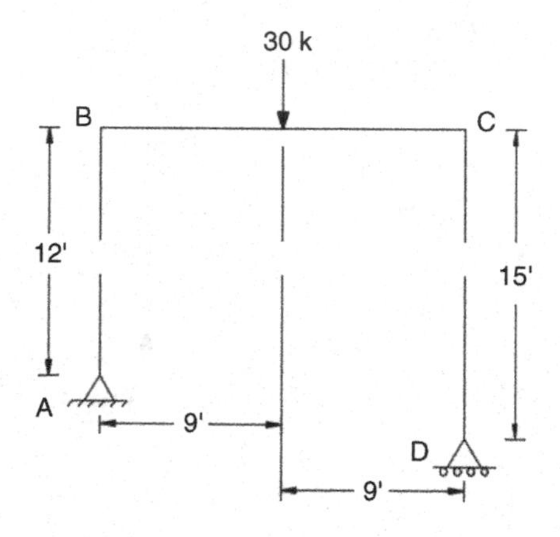

Problem 2.5 Find deflection at point D by using unit load method.

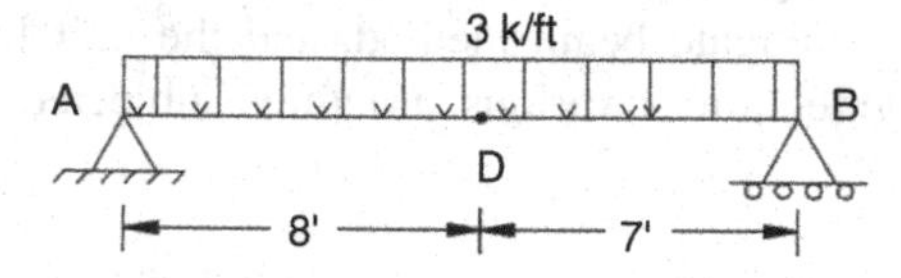

3 Deflection of Statically Determinate Trusses

3.1 BASIC CONCEPT

Trusses are used for various purposes in civil engineering construction. The methods of analysis of a truss vary depending on whether it is a statically determinate or indeterminate truss. This concept is discussed in Section 3.2 and is followed by a few examples.

3.2 STATIC INDETERMINACY (SI)

As pointed out in Section 1.1, the indeterminacy of a truss is determined by the value of SI, calculated using equation 3.1 given below as,

$$SI = b + r - 2j \tag{3.1}$$

where b = number of members in the truss
r = number of support reactions
j = number of joints in the truss

If $SI = 0$, this implies that the truss is determinate. (Except for some cases where a hinge is hanging.)

A few examples are solved to find deflection at specific points in a determinate truss.

3.3 EXAMPLES

Example 3.3.1

Determine the deflection at C along the direction CB for the determinate truss shown in Figure 3.1.

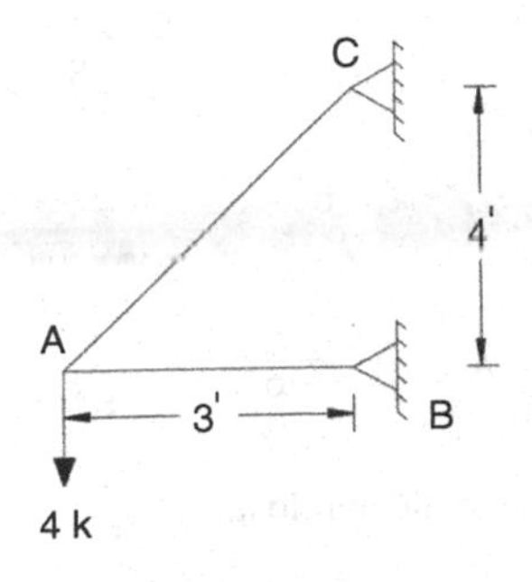

FIGURE 3.1 Determinate Truss under Concentrated Load.

DOI: 10.1201/9781003246633-4

$$SI = b + r - 2j$$

$$SI = 2 + 4 - 2(3) = 0$$

This implies that the truss shown is a determinate truss.
Use

$$\Delta = \Sigma \frac{NnL}{AE}$$

where N = force in a member due to applied load
n = force in a member due to unit load
AE = axial rigidity
L = length of a member

To find deflection at C along CB, the calculation of "N" due to the applied load needs to be found.

Use the method of joints and move from one joint to next in such a way that there are not more than two unknowns at each joint. This allows the problem to be solved using only two equations of equilibrium (ΣF_x=0 and ΣF_y=0) as shown in Figure 3.1a.

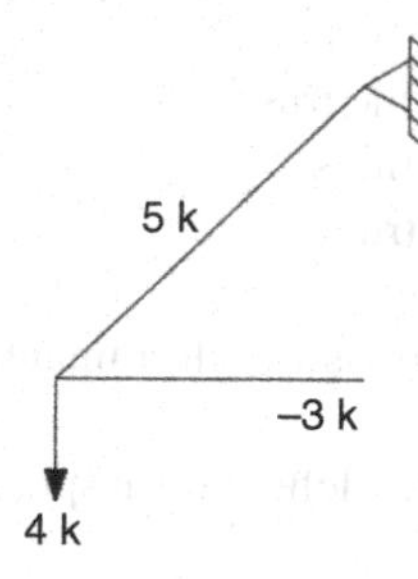

FIGURE 3.1a Values of "N" due to the applied load.

Calculate "n" due to the unit load applied at C along CB after removing the applied load as shown in Figure 3.1b.

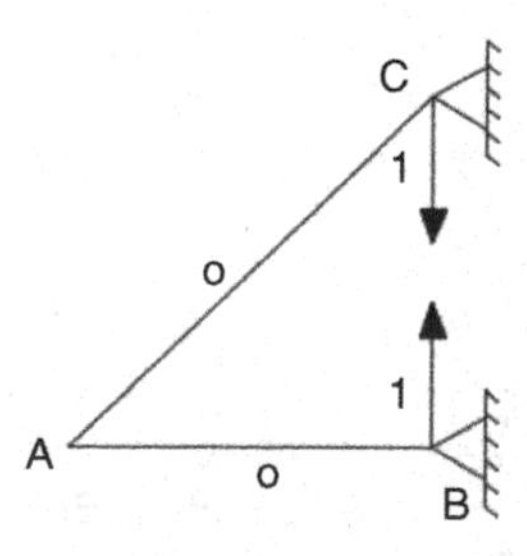

FIGURE 3.1b Values of "n" due to the unit load.

The final "N" and "n" values are shown below.

Member	N	n	L
AB	-3	0	3'
AC	5	0	5'

Use

$$\Delta_{CB} = \Sigma \frac{NnL}{AE}$$

$$\Delta_{CB} = \frac{(0)(-3)(3)}{AE} + \frac{(0)(5)(5)}{AE}$$

$$\Delta_{CB} = 0$$

Example 3.3.2

Find the deflection of joint C along the direction CA for the determinate truss shown in Figure 3.2. Use the unit load method.

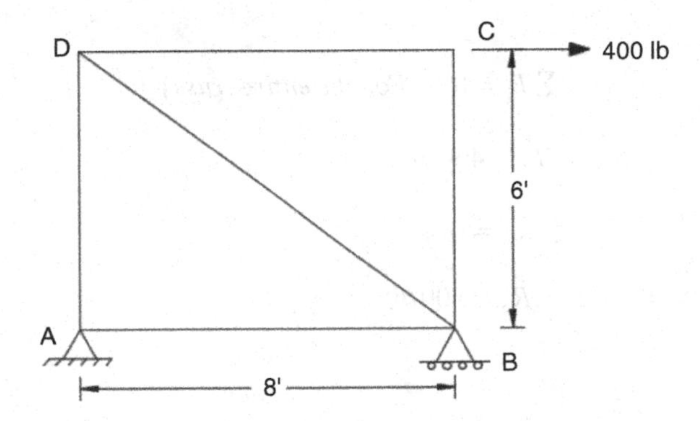

FIGURE 3.2 Determinate Truss under Concentrated Load at C.

$$SI = b + r - 2j$$

$$SI = 5 + 3 - 2(4) = 0$$

This implies that it is a determinate truss.
The basic equation to be used is

$$\Delta = \Sigma \frac{NnL}{AE}$$

The next step is to find the force "N" due to the applied loads, as shown in Figure 3.2a.

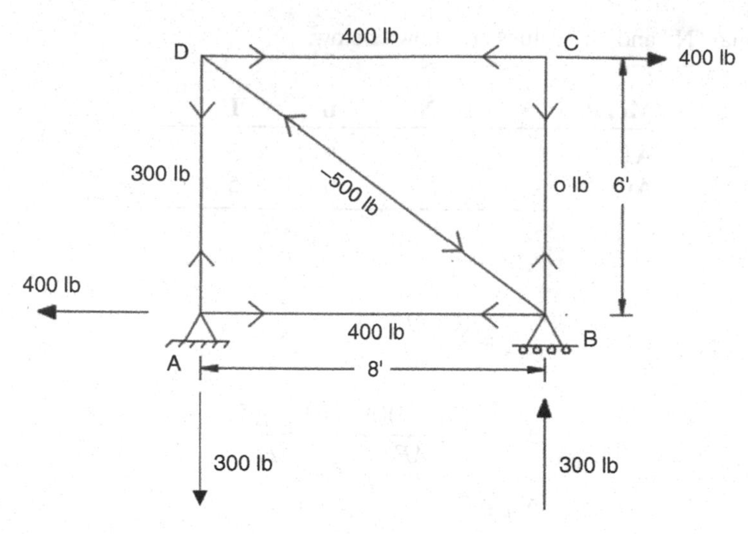

FIGURE 3.2a Values of "N" due to the applied load.

Use the method of joints and move from one joint to next in such a way that there are not more than two unknowns at each joint. This allows the problem to be solved using only two equations of equilibrium ($\Sigma F_x = 0$ and $\Sigma F_y = 0$) as shown in Figure 3.1a.

$$\Sigma F_x = 0 \quad (\textit{For the entire truss})$$

$$H_A = 400\,lb$$

$$\Sigma M_A = 0$$

$$R_B = 300\,lb$$

FBD of joint A

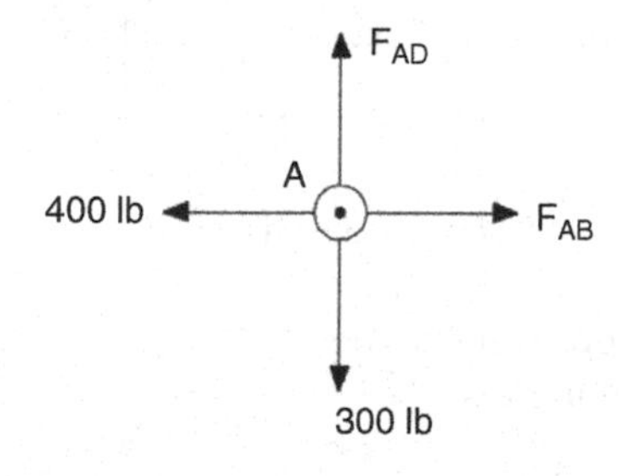

Using $\Sigma F_x = 0$

$$\Sigma F_x = F_{AB} - 400 = 0 \quad \Rightarrow \quad F_{AB} = 400\,lb$$

Using $\Sigma F_y = 0$

$$\Sigma F_y = F_{AD} - 300 = 0 \quad \Rightarrow \quad F_{AD} = 300\,lb$$

FBD of joint D

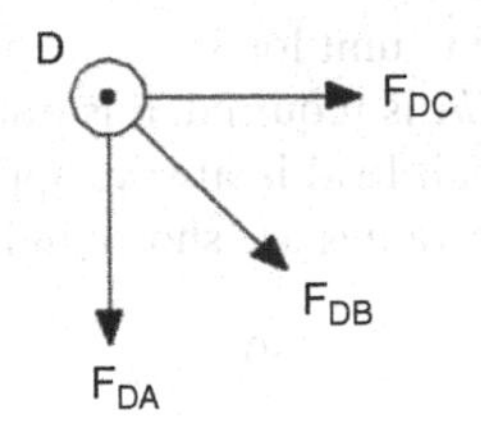

Using $\Sigma F_x = 0$

$$\Sigma F_x = F_{DC} + \left(\frac{4}{5}\right) F_{DB} = 0$$

Using $\Sigma F_y = 0$

$$\Sigma F_y = -F_{DA} - \left(\frac{3}{5}\right) F_{DB} = -300 - \left(\frac{3}{5}\right) F_{DB} = 0$$

Solving these equations gives the following:

$$F_{DB} = -500\,lb$$

$$F_{DC} = 400\,lb$$

FBD of joint C

C

FCD 400 lb

FCB

Using $\Sigma F_x = 0$

$$\Sigma F_x = -F_{CD} + 400 = 0$$

$$F_{CD} = 400\,lb$$

Using $\Sigma F_y = 0$

$$\Sigma F_y = -F_{CB} = 0$$

$$F_{CB} = 0\,lb$$

Calculation of forces "n" due to unit load.

Since deflection of C along CA is required, this implies that a unit load from C to A needs to be applied. As this unit load is always applied along the direction of the required deflection, the resulting values are shown in Figure 3.2b.

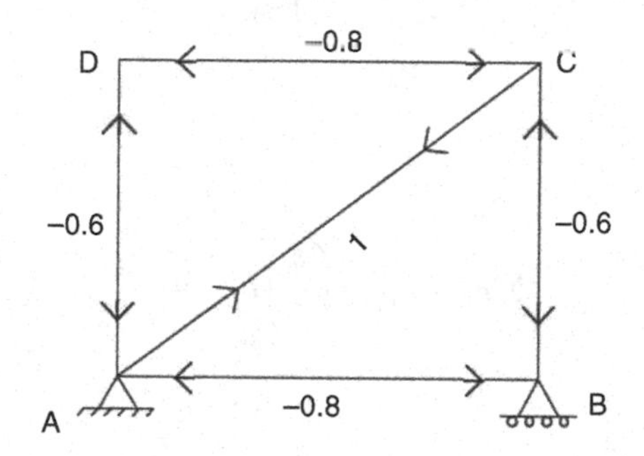

FIGURE 3.2b Values of "n" due to the unit load.

FBD of joint C

C
F_{CD}
$F_{CA} = 1$
F_{CD} F_{CB}

Using $\Sigma F_x = 0$

$$\Sigma F_x = -F_{CD} - \left(\frac{4}{5}\right)F_{CA} = -F_{CD} - 1\left(\frac{4}{5}\right) = 0$$

$$F_{CD} = -0.8$$

Using $\Sigma F_y = 0$

$$\Sigma F_y = -F_{CB} - \left(\frac{3}{5}\right)F_{CA} = -F_{CB} - 1\left(\frac{3}{5}\right) = 0$$

$$F_{CB} = -0.6$$

FBD of joint B

Using $\Sigma F_x = 0$

$$\Sigma F_x = -F_{BA} - \left(\frac{4}{5}\right) F_{BD} = -F_{BA} - 1\left(\frac{4}{5}\right)$$

$$F_{BA} = -0.8$$

Using $\Sigma F_y = 0$ (This was solved in the previous example, so it's not needed.)

$$\Sigma F_x = F_{BC} + \left(\frac{3}{5}\right) F_{BD} = -0.6 + 1\left(\frac{3}{5}\right) = 0$$

FBD of joint A

Using $\Sigma F_y = 0$

$$\Sigma F_y = F_{AD} + \left(\frac{3}{5}\right) F_{AC} = F_{AD} + 1\left(\frac{3}{5}\right) = 0$$

$$F_{AD} = -0.6$$

The final "N" and "n" values are shown below.

Member	N (lb)	n	L (ft)
AB	400	-0.8	8'
BC	0	-0.6	6'
CD	400	-0.8	8'
DA	300	-0.6	6'
DB	-500	1	10'

Now use

$$\Delta = \Sigma \frac{NnL}{AE}$$

$$\Delta_{CA} = \Sigma \frac{\frac{400}{1000} \times (-0.8) \times 8}{AE} + \frac{0 \times (-0.6) \times 6}{AE} + \frac{\frac{400}{1000} \times (-0.8) \times 8}{AE}$$

$$+ \frac{\frac{300}{1000} \times (-0.6) \times 6}{AE} + \frac{\left(\frac{-500}{1000}\right) \times 1 \times 10}{AE}$$

$$\Delta_{CA} = -\frac{11.2}{AE} \quad (ft)$$

3.4 SUMMARY

In this chapter, the concept and application of finding the deflection of a determinate truss are explained using the unit load method. A few examples are then solved to help further explain the concept.

PROBLEMS FOR CHAPTER 3

Problem 3.1 Determine the horizontal and vertical deflection of all the lower joints due to the applied loading shown in Figure 3.3. Use the unit load method.

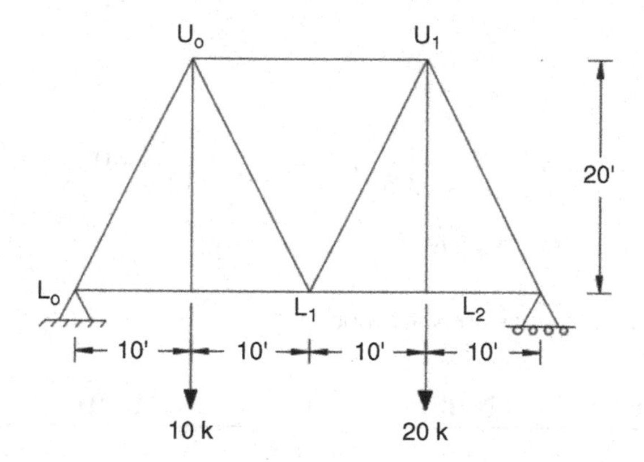

FIGURE 3.3 Determinate truss under two concentrated loads.

Problem 3.2 Determine the vertical displacement of joint C at the steel truss shown in Figure 3.4 due to the applied loading. Use the unit load method.

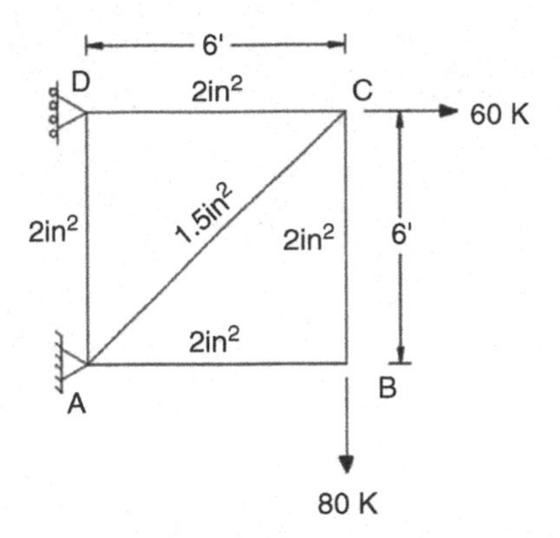

FIGURE 3.4 Determinate cantilever truss under vertical and horizontal concentrated loads.

Problem 3.3 Determine the vertical displacement of joint C of the truss shown in Figure 3.5. $A = 0.5$ in², $E = 29{,}000$ ksi. Use the unit load method.

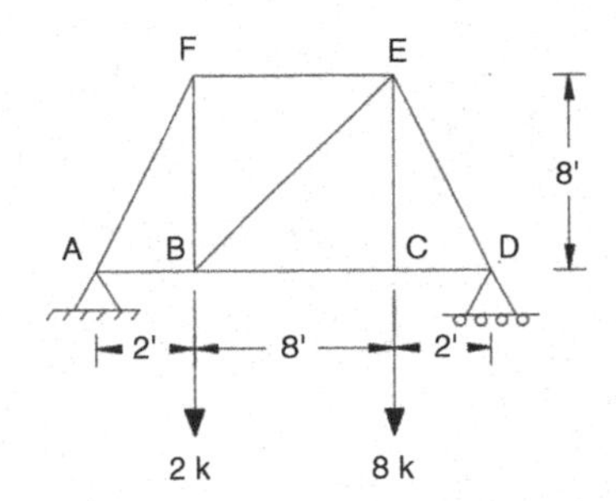

FIGURE 3.5 Determinate truss under two concentrated loads of the truss given below.

Problem 3.4 Determine the horizontal deflection of roller support at B due to the applied load in Figure 3.6. Use the unit load method. The numbers in parenthesis are areas in 50 in². $E = 30{,}000$ ksi.

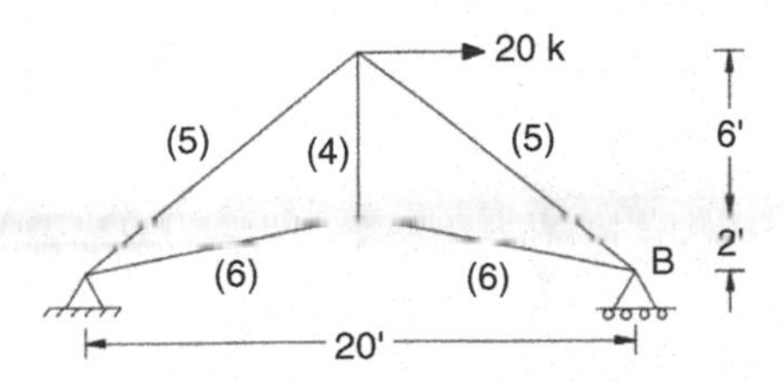

FIGURE 3.6 Simply supported determinate truss under horizontal load.

4 Shear Force and Bending Moment Diagrams for Beams

This chapter deals with the variation of shear force and bending moment along the span of beams. A shear force diagram (SFD) shows the variation of shear force along the span of the beam, while a bending moment diagram (BMD) shows the variation of bending moment along the span of the beam.

To draw an SFD and a BMD, one needs to establish a sign convention as shown in Figure 4.1.

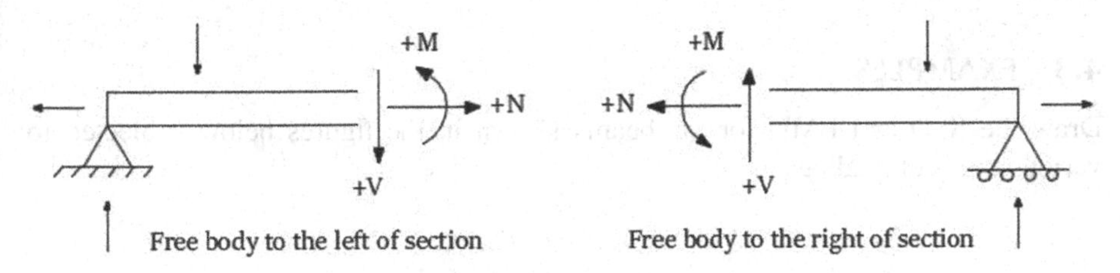

FIGURE 4.1 Concept of shear force (V) and bending moment (M).

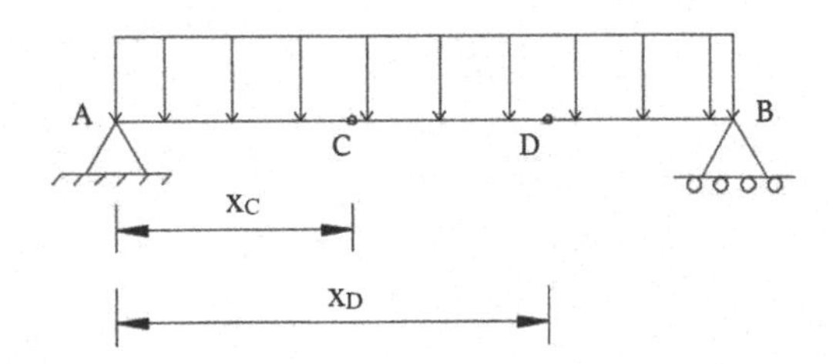

FIGURE 4.2 Simply supported beam acted on by distributed load.

4.1 PROCEDURE TO DRAW SFD AND BMD

1. Find reactions at the supports of the beam using equations of equilibrium (equation 1.1) as explained in Chapter 1.
2. Cut a section at various parts of the beam to the left and right of the portion of the beam wherever the loads are applied.
3. Apply the equations of equilibrium (equation 1.1) again and derive the expressions for shear force (V) and bending moment (M) at various sections of the beam.
4. Using these expressions for V and M as functions of x denotes the distance from the origin along the span of the beam, draw the SFD and BMD.

DOI: 10.1201/9781003246633-5

A few examples for SFD and BMD are shown below using the procedure outlined above.

4.2 RELATION BETWEEN DISTRIBUTED LOAD, SHEAR FORCE, AND BENDING MOMENT

$$V_D - V_c = \int_{x_C}^{x_D} w\,dx \tag{4.1}$$

$$M_D - M_c = \int_{x_C}^{x_D} V\,dx \tag{4.2}$$

where C and D are any two points on the beam and w is the distributed load (see Figure 4.2). It should be noted that if there is no distributed load (w=0) in the entire segment being considered, the slope is zero and the shear force is constant.

It becomes easy to draw SFD and BMD using the above two equations instead of drawing from basic principles.

4.3 EXAMPLES

Draw the SFD and BMD for the beams shown in the figures below subjected to various kinds of loadings.

EXAMPLE 4.3.1

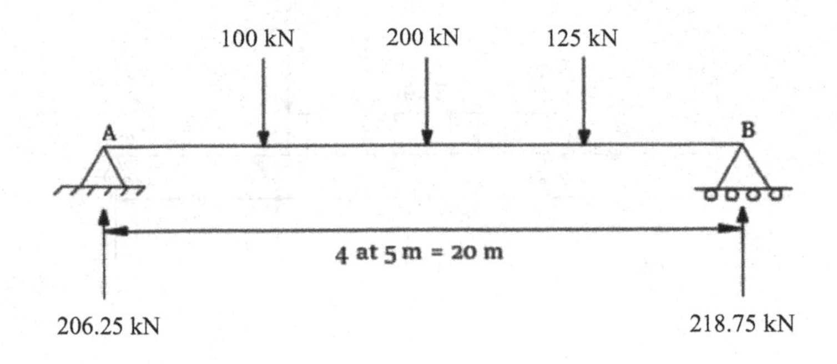
100 kN
200 kN
125 kN
A
B
4 at 5 m = 20 m
206.25 kN
218.75 kN

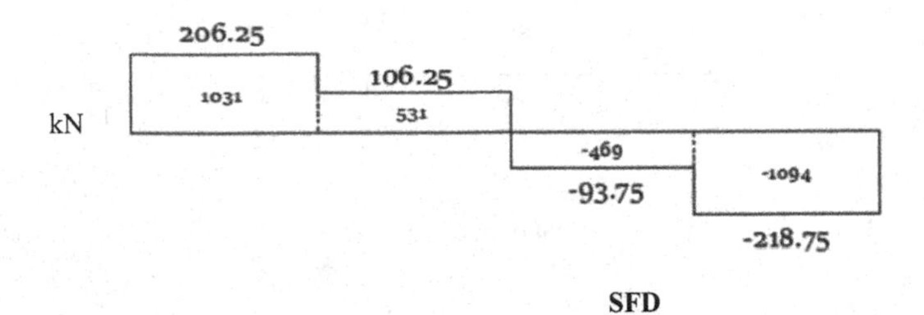
206.25
1031
106.25
531
kN
-469
-93.75
-1094
-218.75
SFD

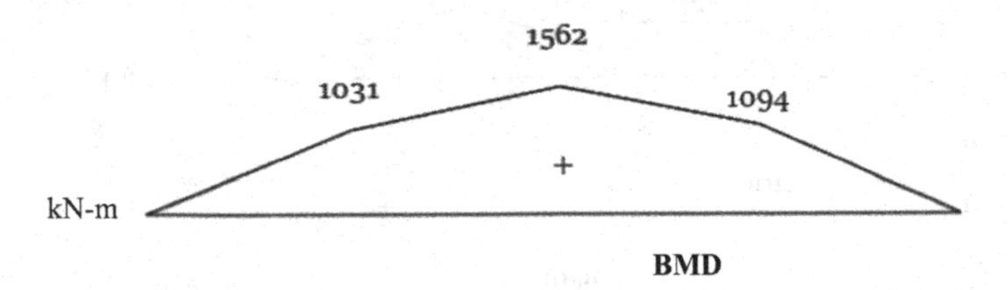
1562
1031
1094
+
kN-m
BMD

Example 4.3.2

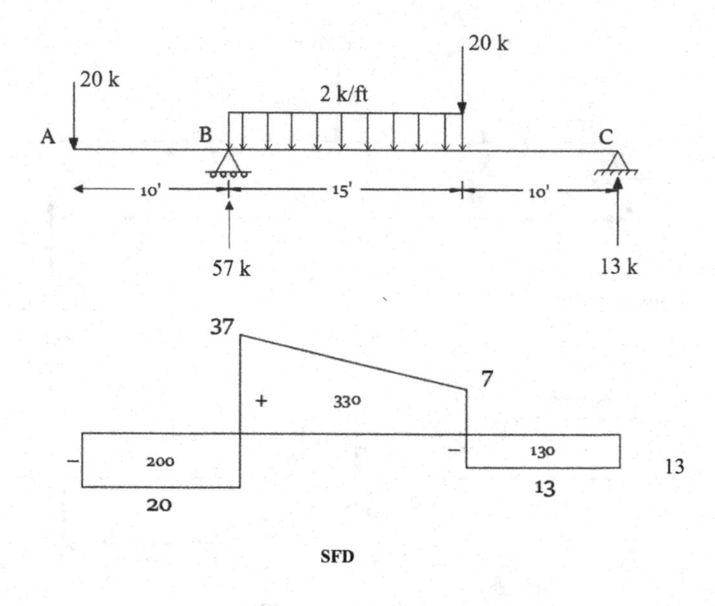
20 k
20 k
2 k/ft
A
B
C
10'
15'
10'
57 k
13 k
37
7
+
330
200
130
13
13
20
SFD

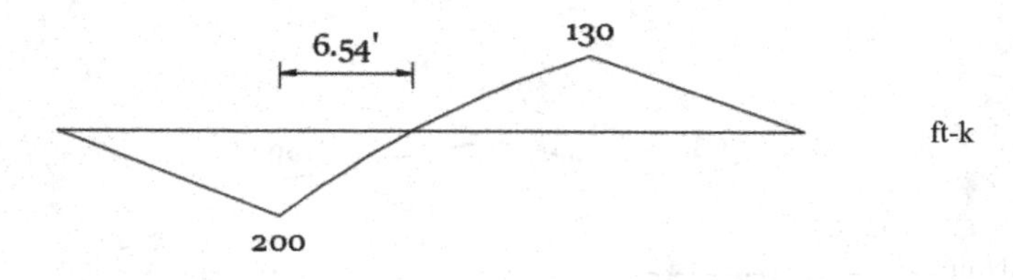
6.54'
130
200
ft-k
BMD

Example 4.3.3

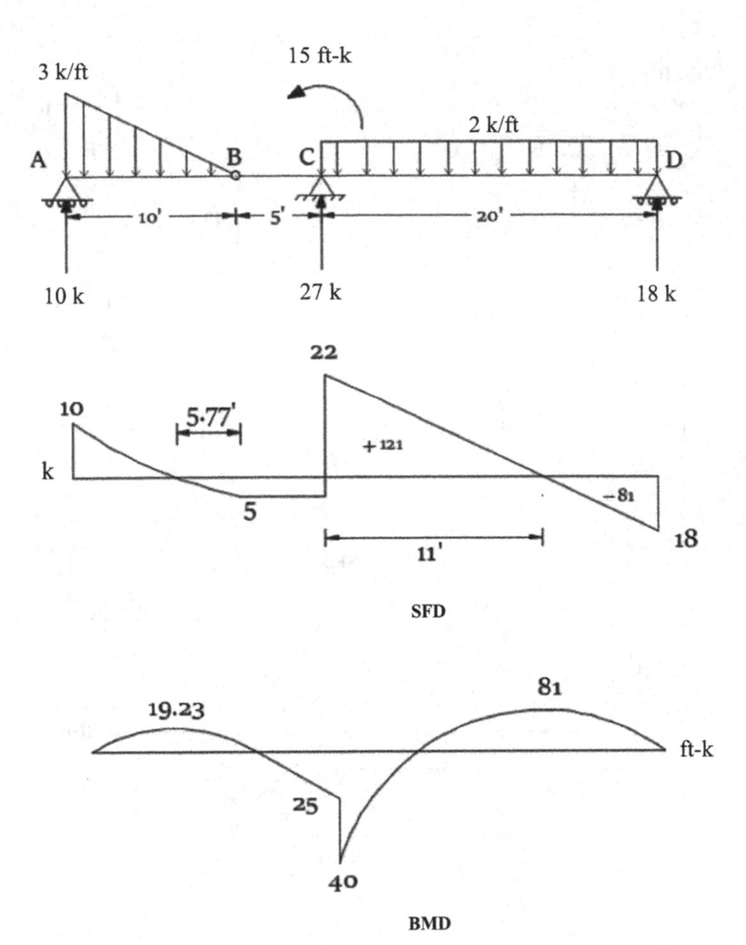
15 ft-k
3 k/ft
2 k/ft
A
B
C
D
10'
5'
20'
10 k
27 k
18 k
22
10
5.77'
+121
k
−81
5
18
11'
SFD
81
19.23
ft-k
25
40
BMD

Example 4.3.4

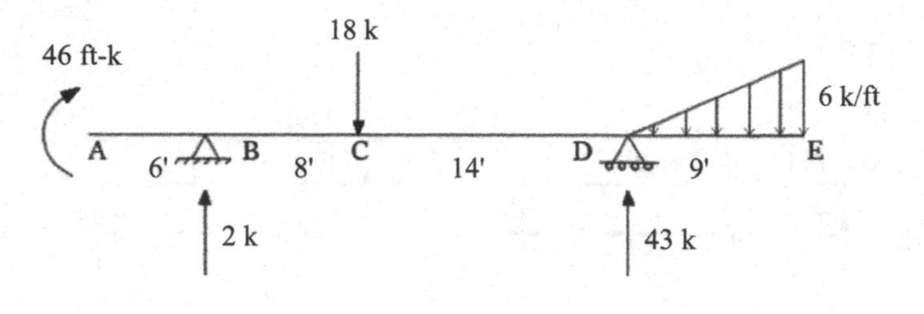

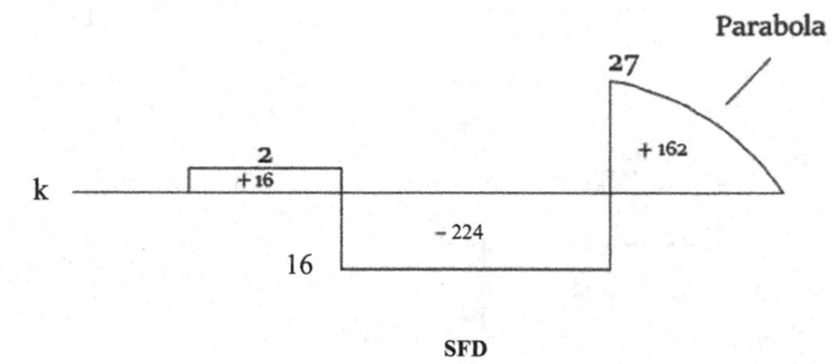

SFD

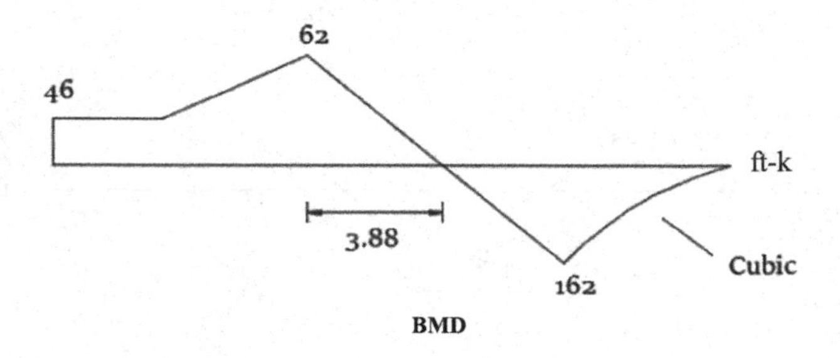

BMD

EXAMPLE 4.3.5

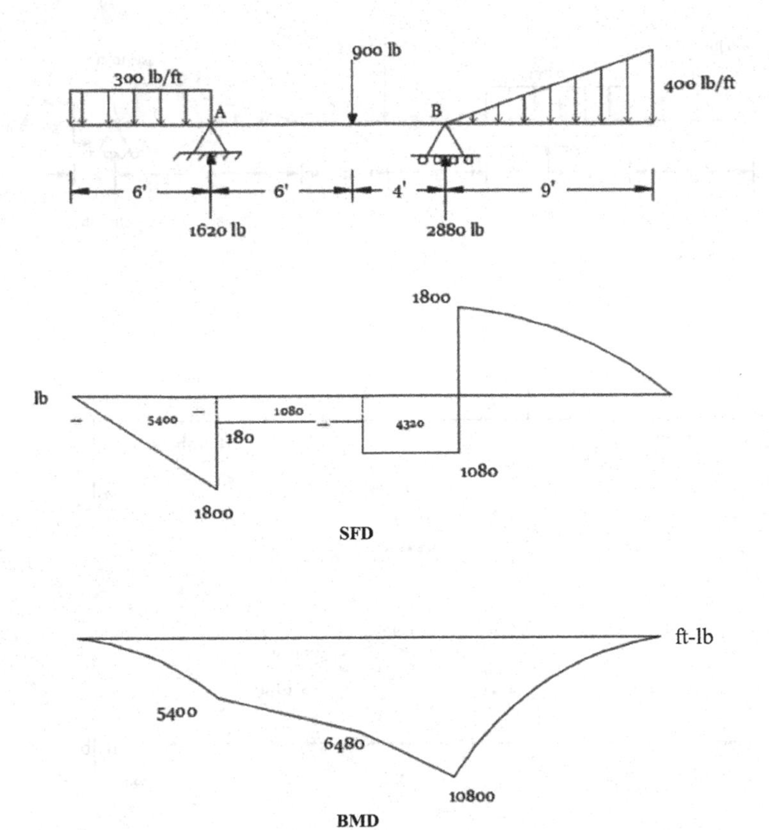
900 lb
300 lb/ft
400 lb/ft
A
B
6'
6'
4'
9'
1620 lb
2880 lb
1800
lb
5400
1080
180
4320
1080
1800
SFD
ft-lb
5400
6480
10800
BMD

EXAMPLE 4.3.6

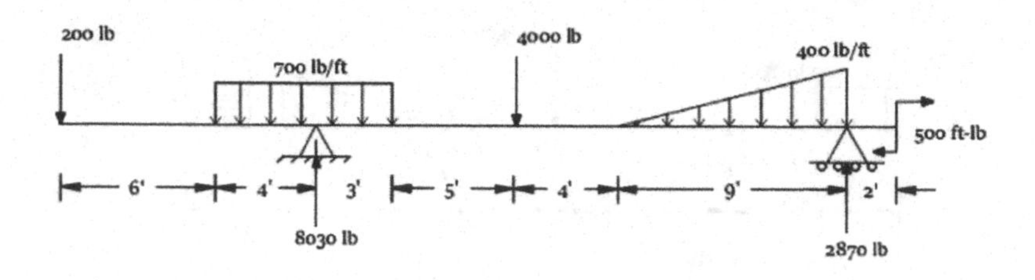

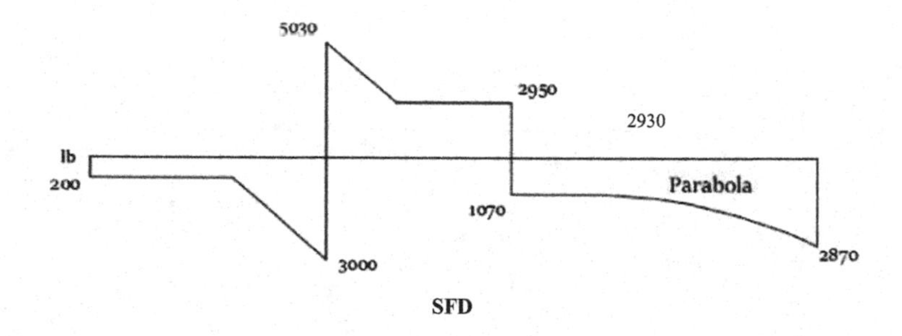

SFD

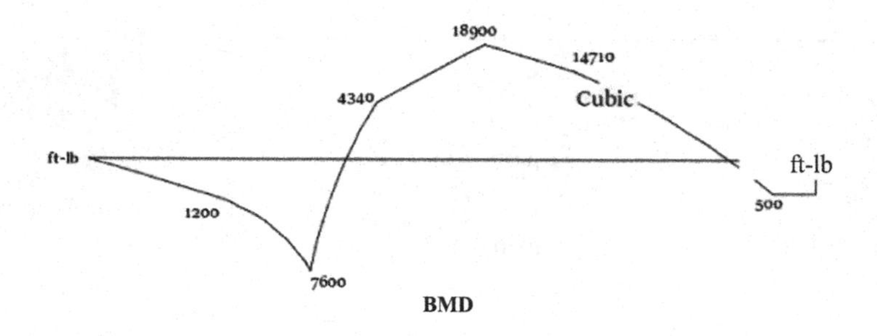

BMD

EXAMPLE 4.3.7

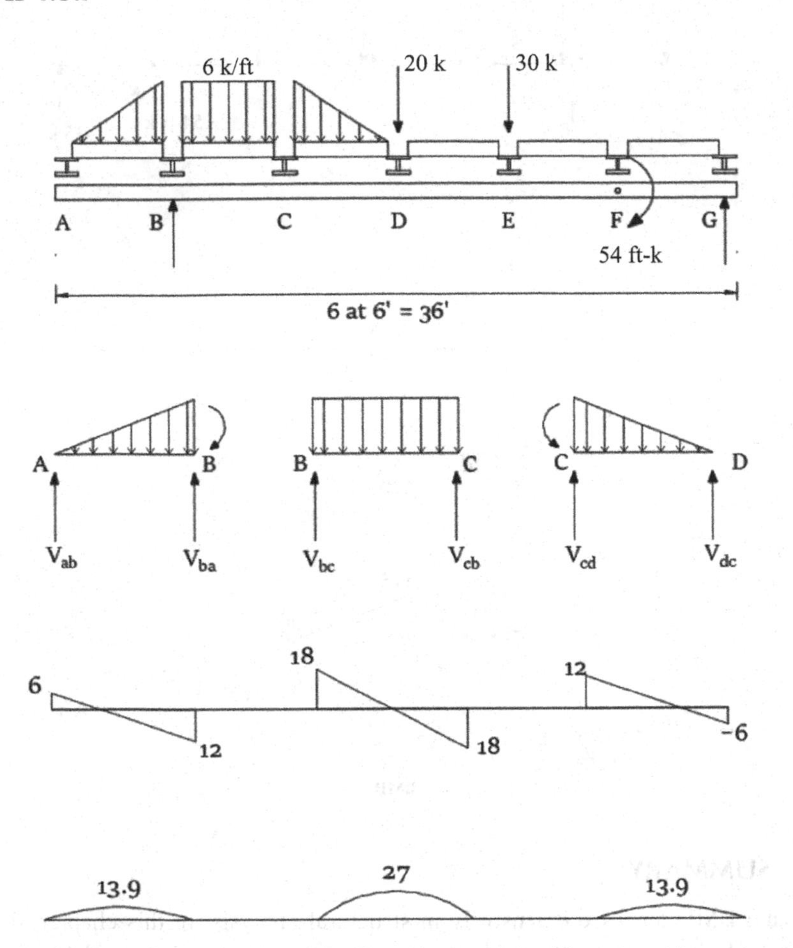
6 k/ft
20 k
30 k
A
B
C
D
E
F
G
54 ft-k
6 at 6' = 36'
A
B
B
C
C
D
V_{ab}
V_{ba}
V_{bc}
V_{cb}
V_{cd}
V_{dc}
6
18
12
12
18
-6
13.9
27
13.9

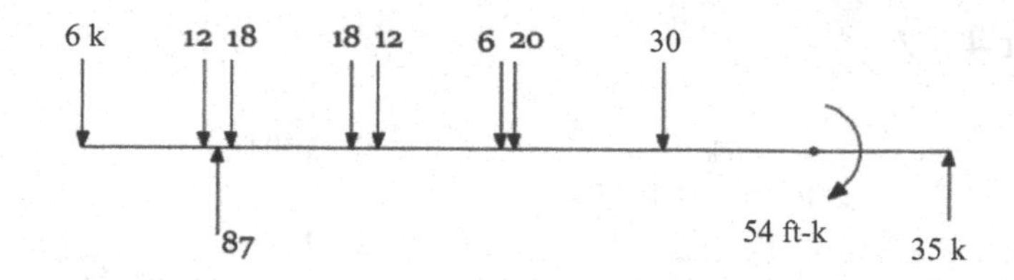

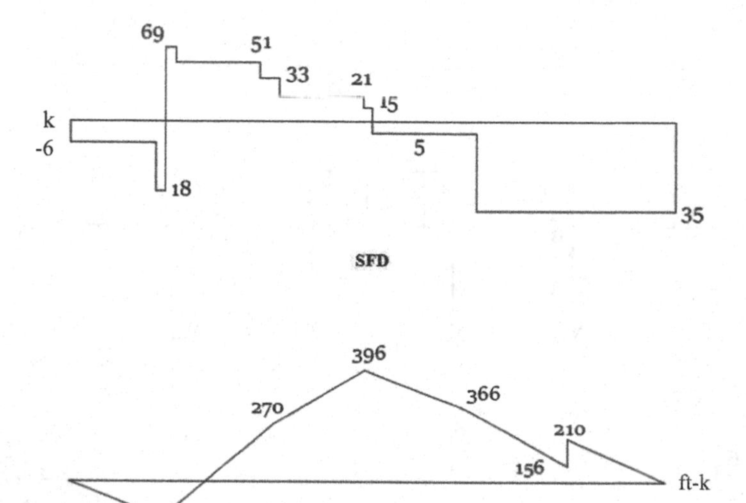

4.4 SUMMARY

SFD and BMD are used extensively in structural analysis. In this chapter, first the concept of shear force and bending moment diagrams are discussed followed by actual examples of how to find them. These concepts are then applied for various beams subjected to different kinds of loadings.

PROBLEMS FOR CHAPTER 4

Problem 4.1 Draw the shear force diagram and the bending moment diagram.

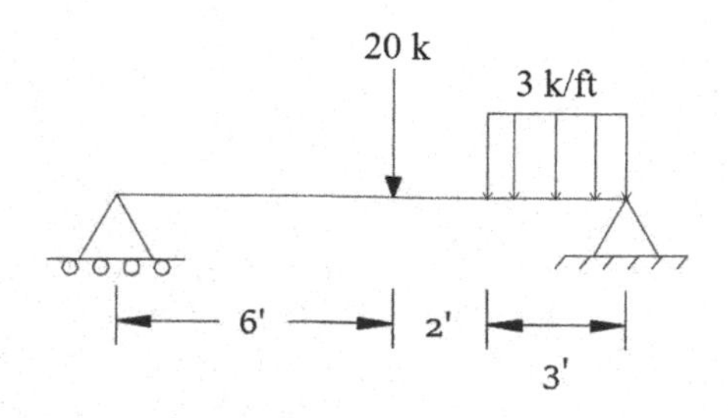

Problem 4.2 Draw the shear force diagram and the bending moment diagram.

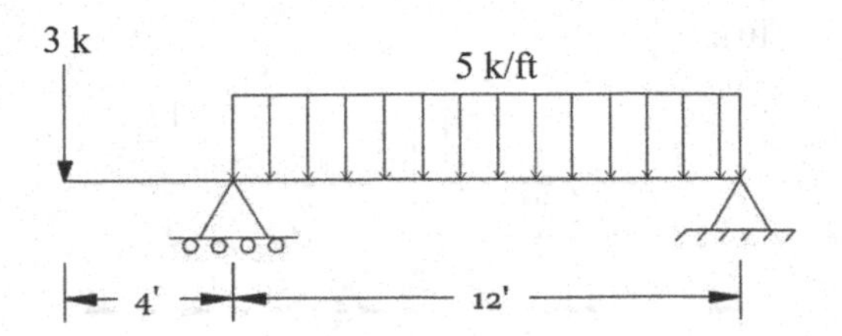

Problem 4.3 Draw the shear force diagram and the bending moment diagram.

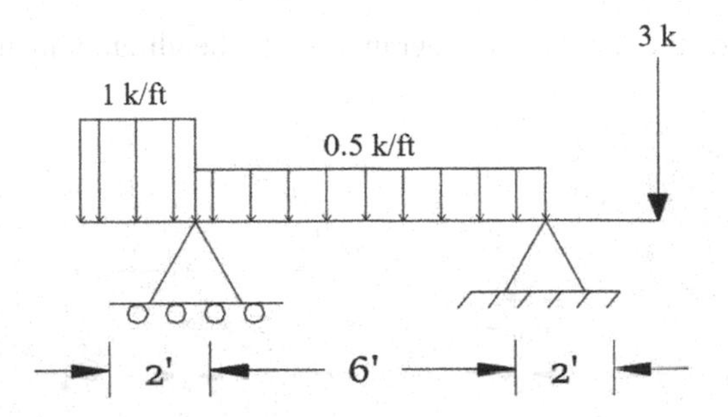

Problem 4.4 Draw the shear force diagram and the bending moment diagram.

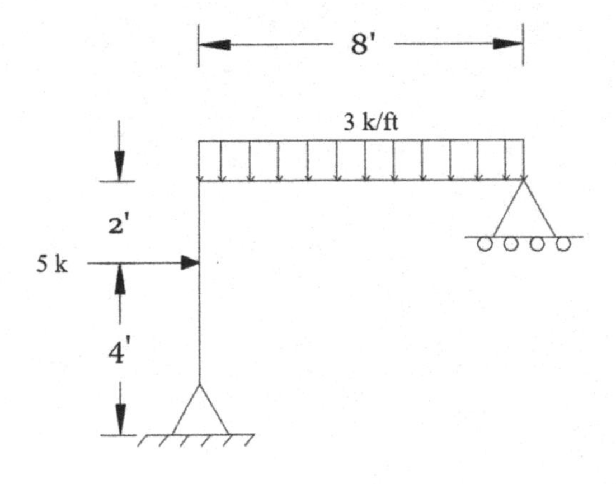

Problem 4.5 Draw the shear force diagram and the bending moment diagram.

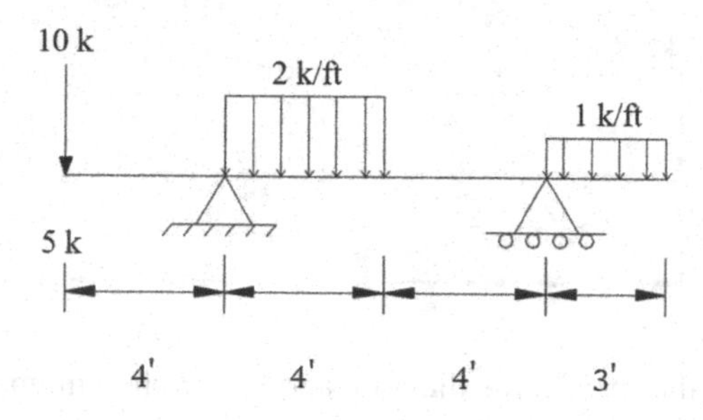

Problem 4.6 Draw the shear force diagram and the bending moment diagram.

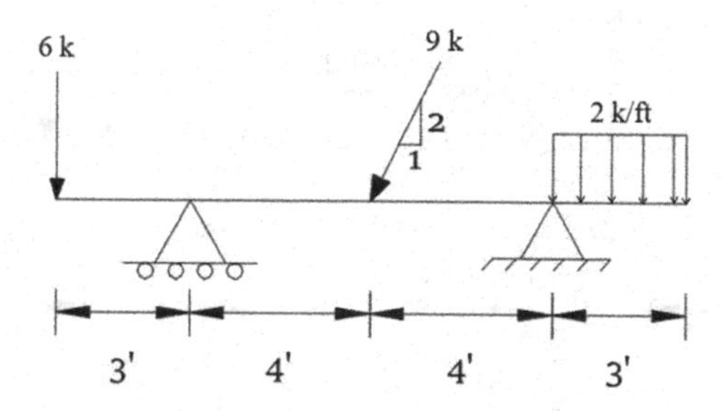

5 Influence Lines for Statically Determinate Structures

5.1 BASIC CONCEPT

A bridge can be considered as a series of simple spans (each one can be considered as a determinate beam) connected through supports. The loads coming on the bridge are moving loads. Certain location of these loads causes maximum shear, reaction, and bending moment at various locations of the bridge. Graphical representations of such variations are called influence line diagrams or simply influence lines. This chapter discusses the influence lines for statically determinate structures.

5.2 INFLUENCE LINES FOR BEAMS

An influence line essentially defines the variation of any force (force or moment) at a given cross section (which is focused while unit load moves) on the structure as the applied load moves from one end of the beam to the other. This force can be reaction, shear, or moment at a point on the beam. Typically, a unit load is considered as a base line so that superposition principles can be applied to scale-up to actual axle loads imparted on a bridge.

5.2.1 Influence Line for Reaction

Consider a simply supported beam as shown in Figure 5.1a.

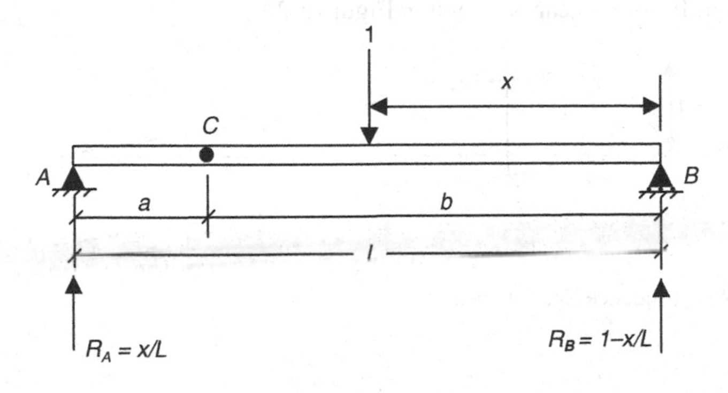

FIGURE 5.1a Influence line for S.S. beam.

DOI: 10.1201/9781003246633-6

The expressions for reaction at A (R_A) and reaction at B (R_B) are given as,

$$R_A = 1 \times \frac{x}{L} = \frac{x}{L} \quad (5.1)$$

$$R_B = 1 - \frac{x}{L} \quad (5.2)$$

The variation of R_A with x along the span of the beam is called the influence line diagram (ILD) for R_A.

The actual influence line diagram for reaction R_A is shown in Figure 5.1b.

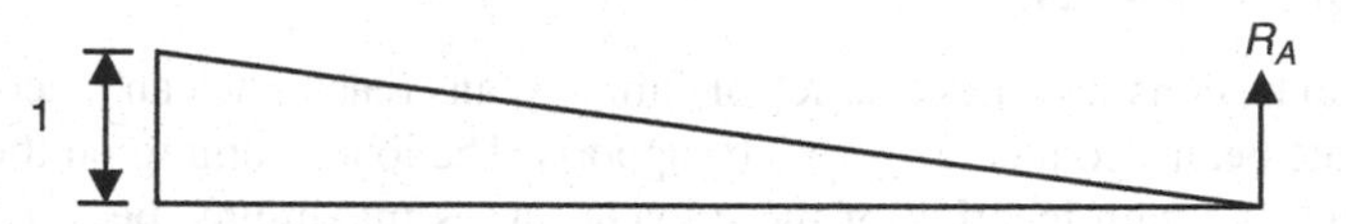

FIGURE 5.1b Influence line for reaction at A.

5.2.2 Influence Line for Shear

The expressions for shear at point C (Figure 5.2a) are given as,

For $0 < x < b$,

$$S_c = R_A = \frac{x}{L} \quad (5.3)$$

For $b < x < L$,

$$S_c = R_A - 1 = -R_B = -\frac{L-x}{L} \quad (5.4)$$

The actual ILD for shear is shown in Figure 5.2a.

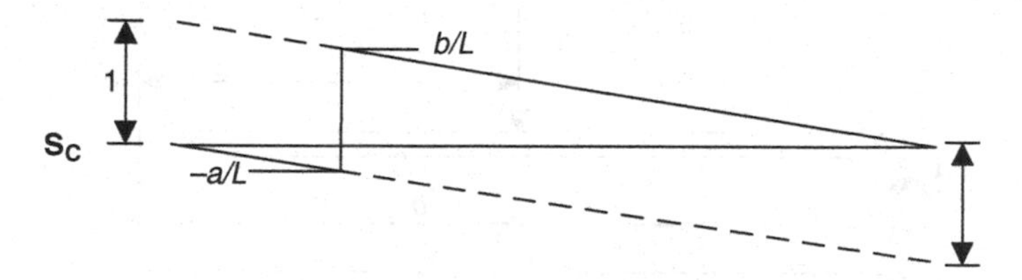

FIGURE 5.2a Influence line for shear at C.

5.2.3 Influence Line for Moment

The expressions for moment at any section of the beam shown in Figure 5.2b are given below:

$For\, 0 \le x \le b,$

$$M_c = R_A \times a = \frac{ax}{L} \tag{5.5}$$

$For\, b \le x \le L,$

$$M_c = R_A \times a - (x-b) = R_B \times b = \frac{b(L-x)}{L} \tag{5.6}$$

The actual ILD for moment is shown in Figure 5.2b.

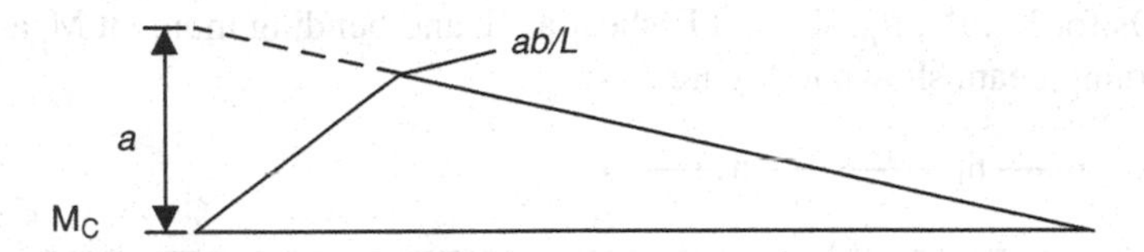

FIGURE 5.2b Influence line for moment at C.

5.3 EXAMPLES

Example 5.3.1

Construct ILDs for reaction R_B and shear at a section just to the right of support B for the beam shown in Figure 5.3a.

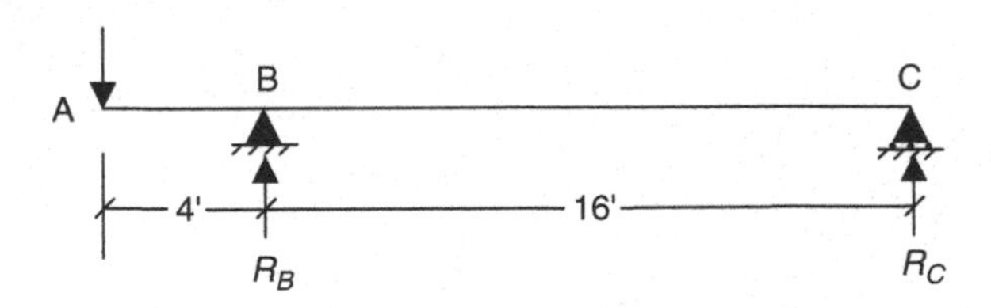

FIGURE 5.3a Diagram of simply supported beam with overhang.

The expressions for R_B and V_B are given as,

$$R_B = \frac{20-x}{16} \tag{5.7}$$

$$V_B = \frac{20-x}{16} \tag{5.8}$$

The actual ILD for R_B is in Figure 5.3b. and for shear just to right of support B is shown in Figure 5.3c.

I.L. for R_B for Simply Supported Beam with Overhang

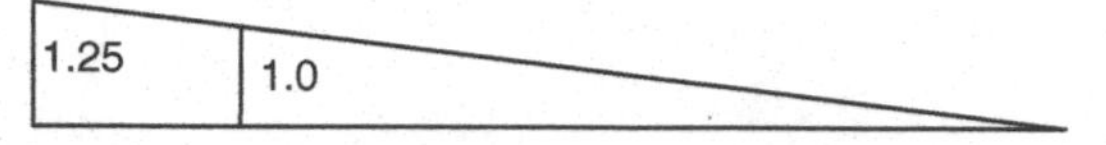

I.L. for shear to the right of R_B for Simply Supported Beam with Overhang

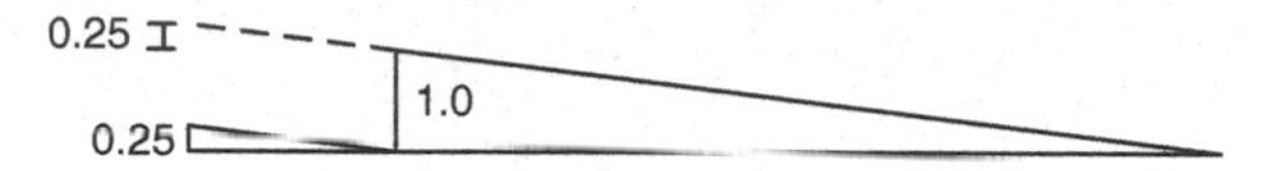

FIGURE 5.3b Influence Lines for R_B and shear to the Right of R_B.

Example 5.3.2

Draw ILDs for R_A, R_B, R_C, shear at F, shear at G, and bending moment M_f and M_g for the compound beam shown in Figure 5.4a.

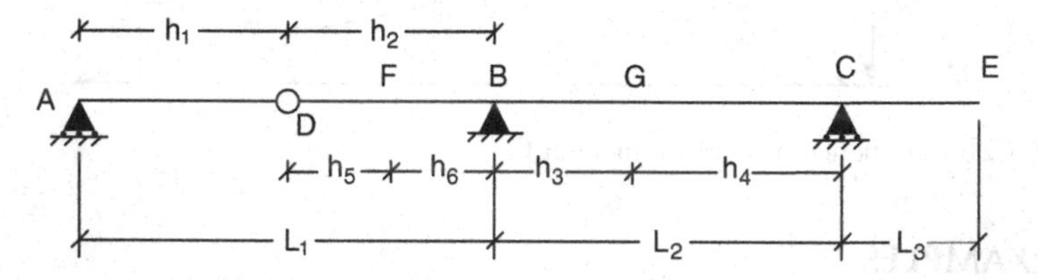

FIGURE 5.4a Diagram of two-span beam with hinge in the left span.

The ILDs are shown in Figure 5.4b.

I.L. for R_A for Two Span Beam with Hinge in Left Span

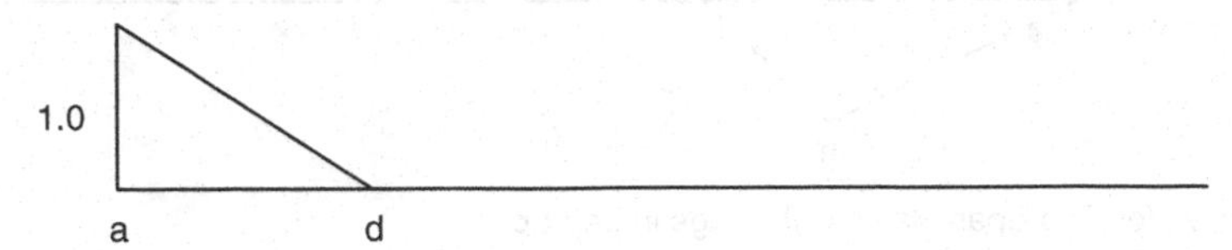

I.L. for R_B for Two Span Beam with Hinge in Left Span

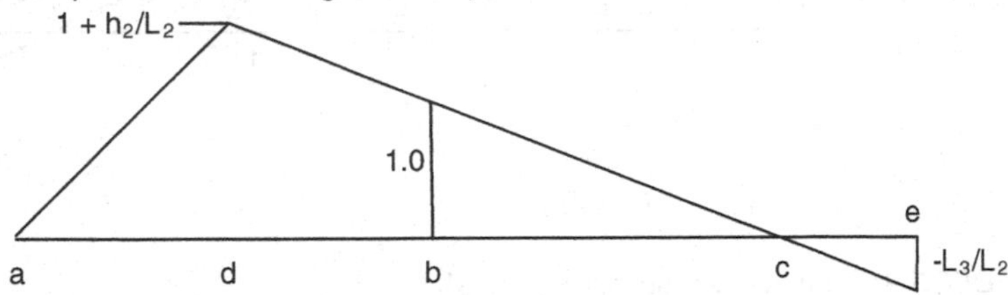

I.L. for R_C for Two Span Beam with Hinge in Left Span

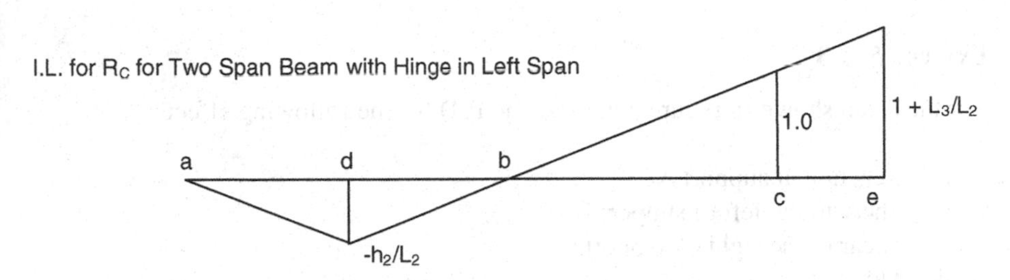

I.L. for shear at f for Two Span Beam with Hinge in Left Span

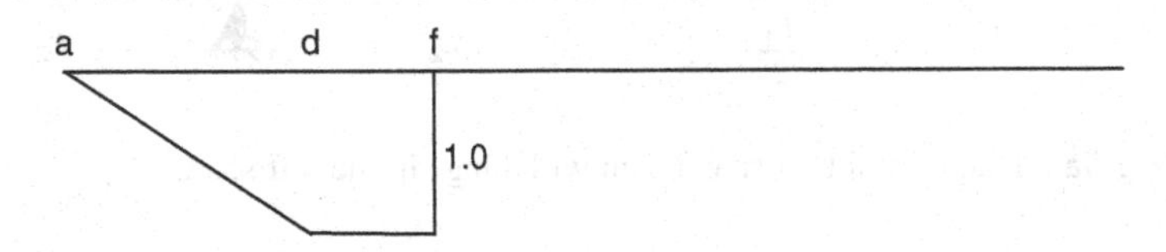

I.L. for shear at g for Two Span Beam with Hinge in Left Span

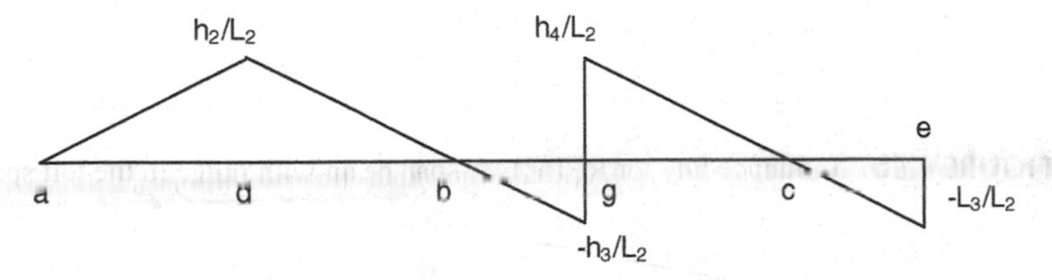

FIGURE 5.4b Influence Lines Diagram for R_A, R_B, R_c, shear at f and g.

I.L. for M_f for Two Span Beam with Hinge in Left Span

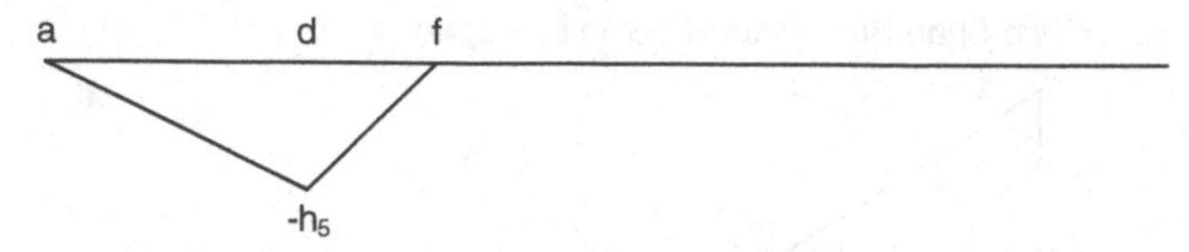

I.L. for M_g for Two Span Beam with Hinge in Left Span

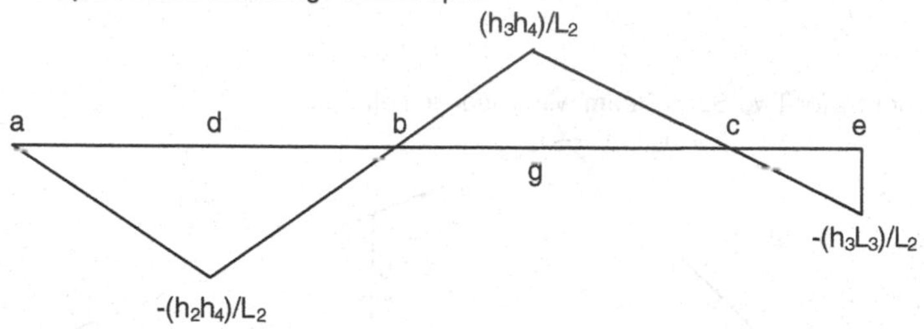

FIGURE 5.4c Influence Lines for M_f and M_g.

Example 5.3.3

For the beam shown in Figure 5.5, draw the ILD for the following effects:

a. Reaction at support A
b. Shear to the left of support B
c. Shear to the right of support C
d. Moment at F

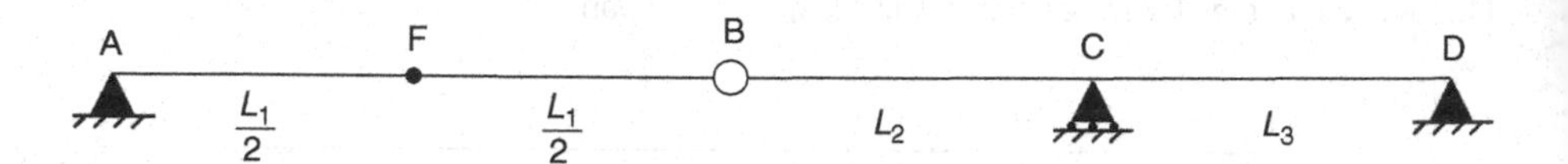

FIGURE 5.5a Diagram of two-span beam with hinge in the left span.

The ILDs are shown in Figure 5.5b–e.

FIGURE 5.5b Influence line for R_A for two-span beam with hinge in the left span.

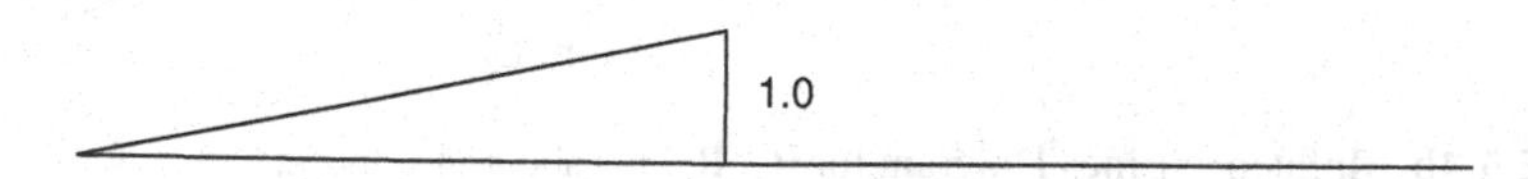

FIGURE 5.5c Influence line for shear to the left of support B for two-span beam with hinge in the left span.

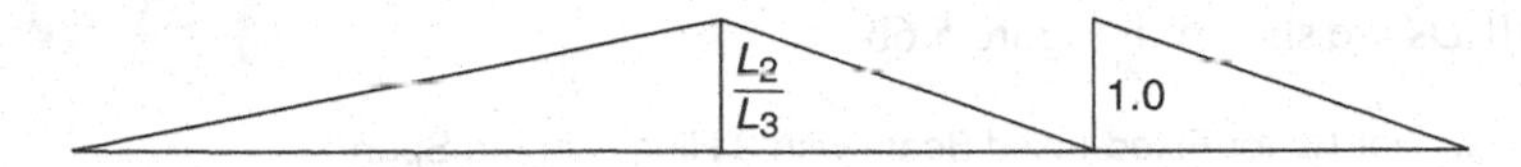

FIGURE 5.5d Influence line for shear to the right of support C for two-span beam with hinge in the left span.

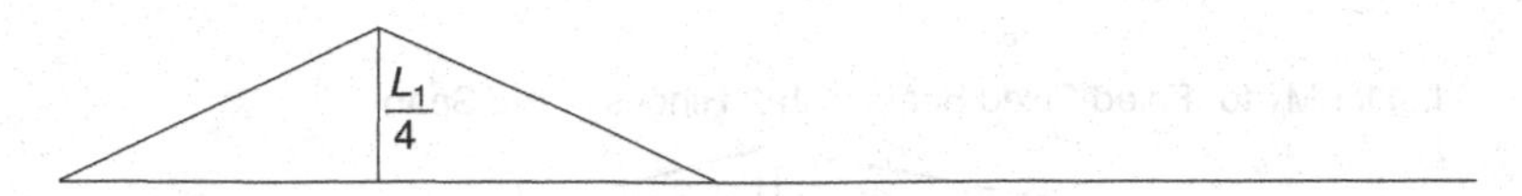

FIGURE 5.5e Influence line for M_F for two-span beam with hinge in the left span.

Example 5.3.4

For the beam shown in Figure 5.6, draw the ILD for the following effects:

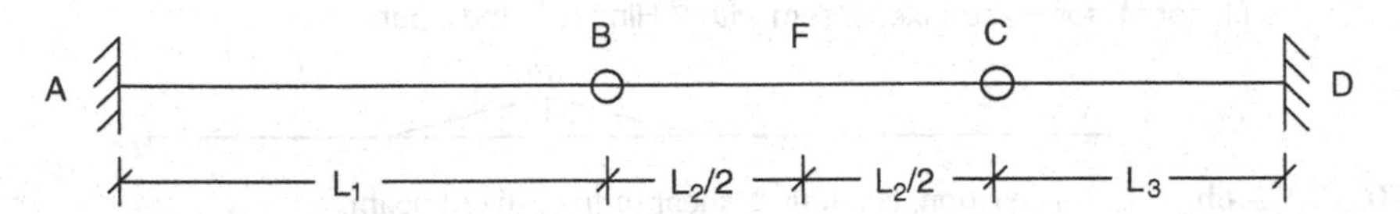

FIGURE 5.6a Fixed-fixed beam with two hinges in the span.

a. Reaction at A
b. Moment at A
c. Shear at C
d. Moment at F

(Assume that no horizontal load exists on the structure.)

The ILDs are shown in Figure 5.6b–e.

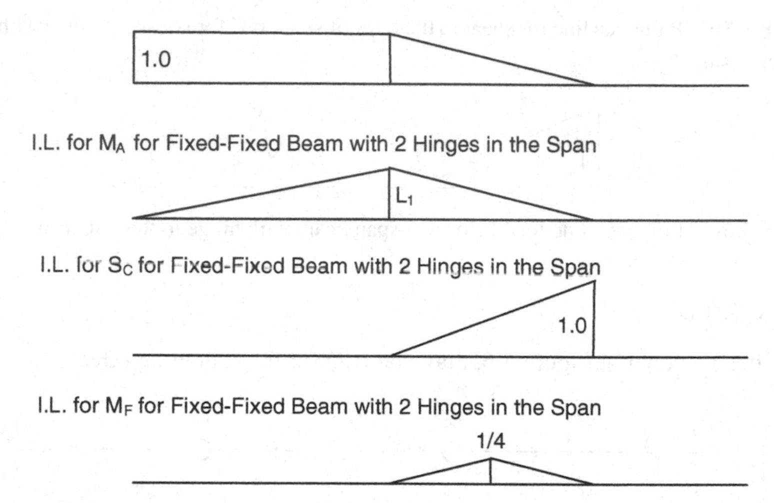

FIGURE 5.6b I,L, for reaction, bending & shear of fixed-fixed beam.

EXAMPLE 5.3.5

For the overhanging beam shown in Figure 5.7, draw the ILD for the following effects:

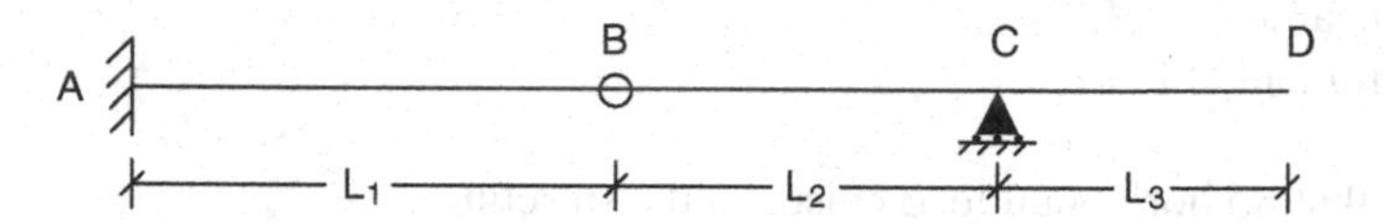

FIGURE 5.7a Fixed-simply supported beam with a hinge in main span.

a. Reaction at A
b. Shear at A
c. Moment at A
d. Reaction at C
e. Moment at C
f. Moment at D

The ILDs are shown in Figure 5.7b.

I.L. for R_A and S_A for Fixed-Simply Supported Beam with a Hinge in Main Span

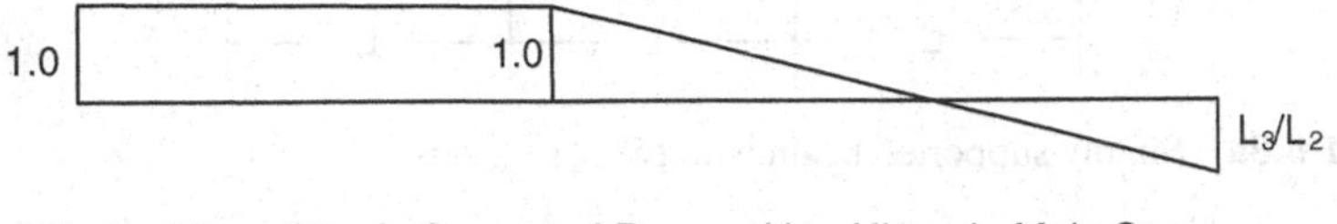

I.L. for M_A for Fixed-Simply Supported Beam with a Hinge in Main Span

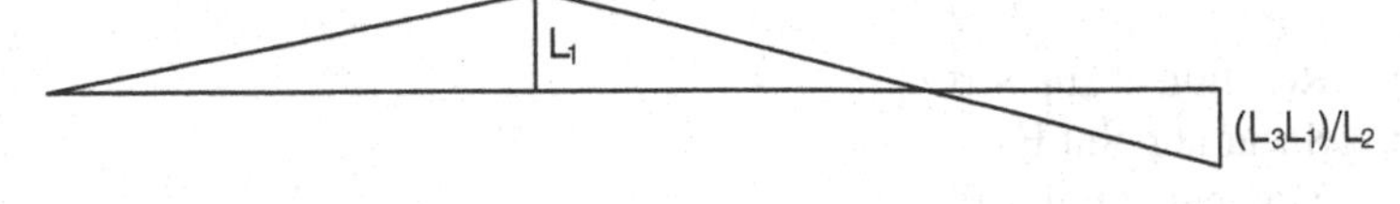

I.L. for R_C for Fixed-Simply Supported Beam with a Hinge in Main Span

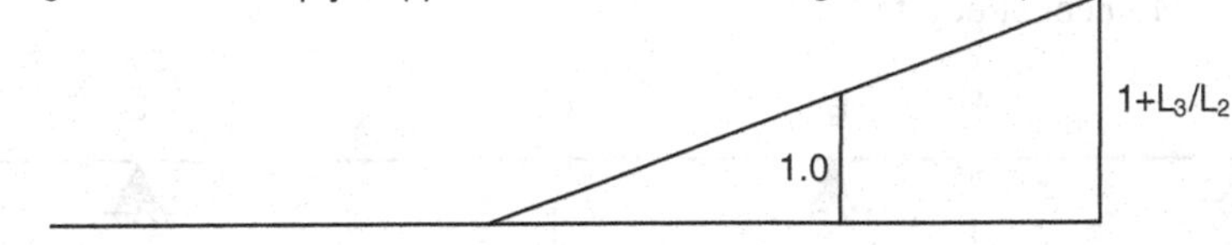

I.L. for M_C for Fixed-Simply Supported Beam with a Hinge in Main Span

I.L. for M_D for Fixed-Simply Supported Beam with a Hinge in Main Span

FIGURE 5.7b I.L. reaction and moment of beam with overhang.

PROBLEMS FOR CHAPTER 5

Problem 5.1 For the beam shown below, draw the influence line diagrams (ILDs) for the following effects:

a. Reaction at support C
b. Shear at section just to the left of support C
c. Shear at section just to the right of support C
d. Moment at point B
e. Moment at point C
f. Moment at point D

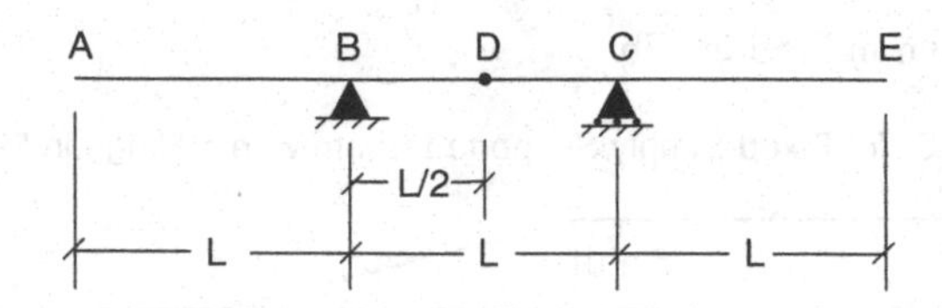

FIGURE 5.8a Simply supported beam with two overhangs.

Problem 5.2 For the beam shown below, draw the ILDs for the following effects:

a. Reaction at support B
b. Shear at point F
c. Moment at point B
d. Moment at point C

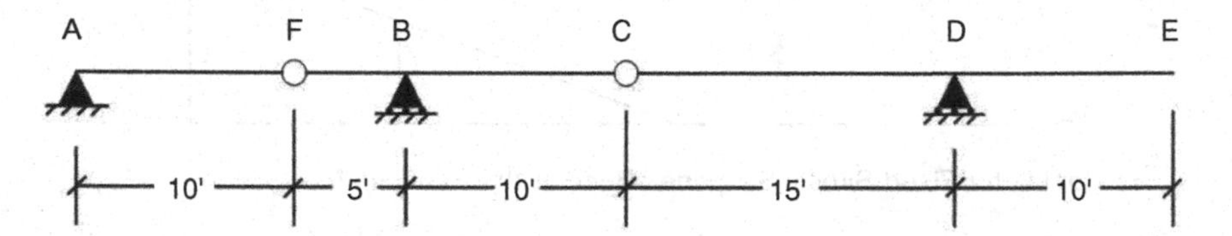

FIGURE 5.8b Two-span beam with overhang and two hinges between each support.

Part II

Analysis of Statically Indeterminate Structures

6 Analysis of Statically Indeterminate Structures by the Force Method (Flexibility Method or Method of Consistent Deformation)

6.1 BASIC CONCEPTS OF THE FORCE METHOD

The force method, which is also called the flexibility method or the method of consistent deformations, uses the concept of structural static indeterminacy (SI). The force method becomes cumbersome when the SI of a structure is large. The results obtained by solving the problem using the force method are all the unknown forces (such as reactions at the supports).

If one is interested in finding rotational or translational displacements of an indeterminate structure, they must be obtained separately using any methods of finding displacements (e.g., unit load method, moment area method or conjugate beam method for example).

This method is applicable to any kind of beam, frame, or truss. It is to be noted that beam and frame structures are predominantly bending (flexure) structures, while trusses are predominantly direct stress structures (tension or compression) in nature. The truss members are not subjected to bending in theory, even though joints freeze and exert small moments on to a member. In other words, all loads are axial.

6.1.1 List of Symbols and Abbreviations Used in the Force Method

Symbols and terms are defined along with equations. However, some are not in equations so they are defined below:

$\left(\Delta_A\right)_L$: Deflection at point A due to applied loading
$\left(\Delta_A\right)_R$: Deflection at A due to redundant loading $\equiv R_A * \delta_{aa}$
α_{aa}: Rotational deflection at A due to a unit load at A
θ_A: Rotational deflection at A due to applied loading
δ_{aa}: Deflection at A due to a unit load at A

DOI: 10.1201/9781003246633-8

6.2 STATIC INDETERMINACY

The SI for beams and frames is defined as,

$$SI = n_u - n_e \tag{6.1}$$

where n_u = Number of unknown support reactions
n_e = Number of equations of equilibrium.

It should be noted that equation (6.1) deals with only external indeterminacy. For structures like trusses, where internal indeterminacy is involved, a different equation is used, which will be discussed later in this chapter.

It should also be noted that equation (6.1) is valid for cases with no internal hinges. It should also be noted that when the number of equations of equilibrium (n) is more than the number of unknowns, the structure becomes unstable.

In general, for two dimensional beams and frames, there are three equations of equilibrium ($n_e = 3$) and for three dimensional beams and frames there are six ($n_e = 6$). The SI refers to the number of reactions that are unsolvable using basic statics.

This implies that a structure is statically determinate if $SI = 0$. An example of this would be a simply supported beam (where one end is pinned and the other has a roller support). This structure would have three unknowns: the reactions at the pin in both the x and y directions and the reaction at the roller in the y direction. The number of equilibrium equations would be 3 ($\sum F_x = 0$, $\sum F_y = 0$, and $\sum M = 0$). Therefore, $SI = 0$.

If $SI \geq 1$, then the structure is said to be statically indeterminate to that degree (value of SI); therefore, the degree of SI is equal to the value of $(n_u - n_e)$. It can be also said that the structure has "SI" number of redundants.

To solve a statically indeterminate beam (or frame) using the force method, we will use redundant forces. A *redundant force* is one that cannot be solved using static equilibrium equations alone. The forces will be taken out and reapplied so that the considered structure is always statically determinate. Additionally, the *principle of superposition* is applied and deflection will be found as an intermediate step to solving for a given redundant. This process will be further explained in the text along with examples.

For a plane truss, static indeterminacy involves both external and internal indeterminacy because of the internal members in a truss.

SI in the case of a truss is defined as,

$$SI = b + r - 2j \tag{6.2}$$

where b = Number of members in the truss
r = Number of reactions at the supports
j = Number of joints in the truss.

The analysis of an indeterminate structure is split into a series of determinate structures acted on by applied loads (in the original structure) and acted on by redundant force(s). In both cases, the deflections need to be found. Hence, the unit load method for finding deflection will be discussed briefly for a determinate structure.

6.3 BASIC CONCEPTS OF THE UNIT LOAD METHOD FOR DEFLECTION CALCULATION

The unit load method is also referred to as the method of virtual work. The basic equation to calculate displacement (whether translational or rotational) at a given point of a beam or frame is given as,

$$\Delta = \int \frac{M\,m\,dx}{EI} \tag{6.3}$$

where M = Moment at any point in a structure due to applied loads
m = Moment at any point in a structure due to the unit load (force or moment) at the point of interest corresponding to the parameter of interest (deflection or rotation)
E = Modulus of elasticity
I = Moment of inertia of the cross-section of a member

Note: Equation (6.3) has been derived using energy principles.

To find m, the applied loads are removed and a unit load (force or moment) is applied at the point of interest, or redundancy, in a structure. If one is interested in finding a vertical deflection at a point in a structure, then a unit vertical force is applied (as it corresponds to vertical deflection). If one is interested in finding horizontal deflection at a point, then a unit horizontal force is applied at that point as it corresponds to horizontal deflection. Similarly, rotational displacement at a point in a structure is found by applying a unit moment as it corresponds to rotation.
The basic expression for finding displacement at a given point on a truss is given as,

$$\Delta = \sum \frac{NnL}{AE} \tag{6.4}$$

where N = Force in a truss member due to applied loads
n = Force in a truss member due to unit load applied at the point where the deflection is to be obtained
L = Length of a truss member
A = Cross-sectional area of a truss member

Note: The summation in equation (6.4) includes all truss members.

To find n in a truss, the applied loads are removed and a unit load is applied at the point of interest. For example, if one is interested in finding a horizontal deflection at a point in a truss, a horizontal unit load is applied at that point. If vertical deflection at a point in a truss is of interest, then a unit vertical load is applied at that point. In the force method for a truss, whole members are taken as redundant.

6.4 MAXWELL'S THEOREM OF RECIPROCAL DEFLECTIONS

This theorem states that,

$$\delta_{AB} = \delta_{BA} \tag{6.5}$$

where δ_{AB} = Deflection at A due to a unit load applied at B
δ_{BA} = Deflection at B due to a unit load applied at A

Maxwell's theorem reduces the work needed to solve a statically indeterminate structure as it relieves several computations of deflection. For more details, the reader is advised to read the books by Chajes (1983), Wang (1953), and Hibbeler (2012).

6.5 APPLICATION OF FORCE METHOD TO ANALYSIS OF INDETERMINATE BEAMS

1. Calculate the SI of the structure using equation (6.1) or (6.2) depending on whether the structure is beam, frame, or truss.
2. Choose one of the reaction forces (or internal members of the truss) as the redundant force – one at a time if there are multiple redundancies.
3. Split the statically indeterminate structure into a determinate structure (acted upon by applied loads on the structure) and determinate structure(s) acted on by the redundant forces (one at a time).
4. Analyze the determinate structures by the unit load method to find the displacement Δ_L, which is the displacement for the applied loading and redundant removed. Then find δ, which is the displacement for the unit load only, at the point of redundancy. If a moment is taken as redundant, the corresponding displacements will be θ and α.
5. Finally, formulate equation(s) of displacement compatibility at the support(s) (in the case of beams and frames). In the case of trusses, displacement compatibility of truss bars will be used.
6. Solve these equation(s) to get the redundant force(s).
7. Calculate all the reactions at the supports (in addition to the redundant force already determined in step 6) using principles of statics.

A total of three examples are solved in this chapter: one for a statically indeterminate beam, one for a statically indeterminate frame, and one for statically indeterminate truss. While any of the methods for finding deflection (e.g., double integration, moment area method, conjugate beam method, unit load method, or any other existing method) can be used to find displacements (translation or rotation), the authors recommend the use of the unit load method because it is conceptually straight forward and easy to use.

6.5.1 Sign Convention

The following sign convention will be used for the force method:

- Counter-clockwise moments and displacements are positive.

This is often referred to as the right-hand rule.

When a member undergoes bending:

- Compression on a member's top fiber is positive bending.
- Compression on a member's bottom fiber is negative bending (see Figure 6.1).

a) Positive bending – top fiber compression

b) Negative bending – bottom fiber compression

FIGURE 6.1 Bending sign convention.

6.5.2 Example of an Indeterminate Beam

An example dealing with the analysis of a statically indeterminate beam using the force method is solved below.

Example 6.5.2.1

Determine the reactions at the supports for the statically indeterminate structure shown in Figure 6.2 by the force method. Use R_B as the redundant. Take $E = 29000$ ksi and $I = 446$ in^4.

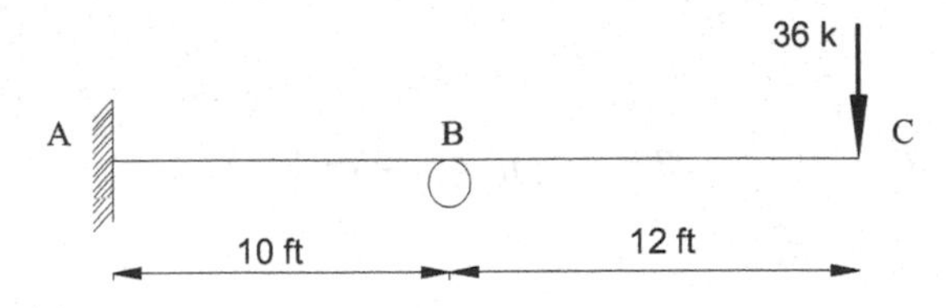

FIGURE 6.2 Statically indeterminate beam.

Solution:

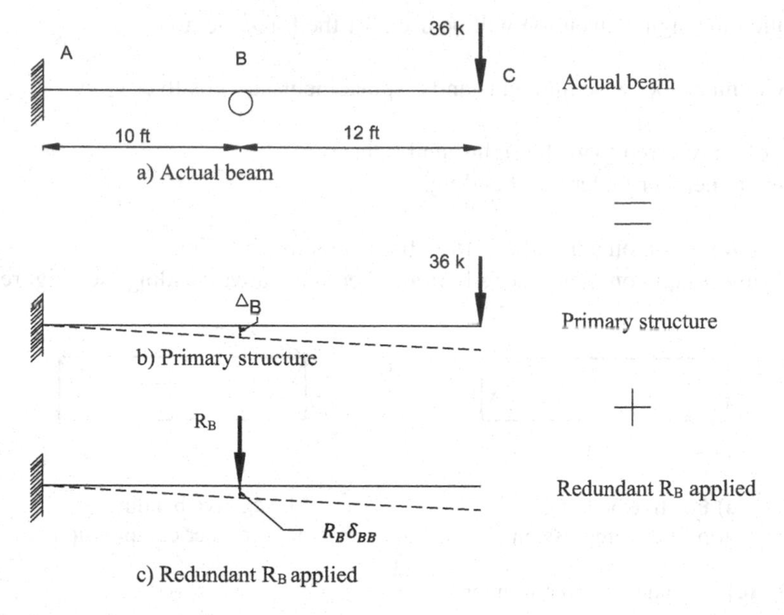

FIGURE 6.3 Two determinate structures.

The given indeterminate structure is split into two determinate structures as shown in Figure 6.3 choosing R_B as the redundant force. The basic equation used is as given in equation (6.3). This is stated again below:

$$\Delta = \int \frac{M\,m\,dx}{EI} \tag{6.3}$$

The procedure will be followed as it is stated earlier in this section.

Step 1

$$SI = n_u - n_e = 4 - 3 = 1$$

Step 2

Choose R_B as the redundant force (in the problem statement).

Step 3

The two determinate structures are shown below with Figure 6.4a acted on by the applied loading, and Figure 6.4c acted on by the redundant force R_B (unit load). The deflection at B for the statically determinate structure $(\Delta_B)_L$ due to applied loads can be obtained from equation (6.3) using the values of M and m. Figures 6.4a–6.4d are used to calculate M and m.

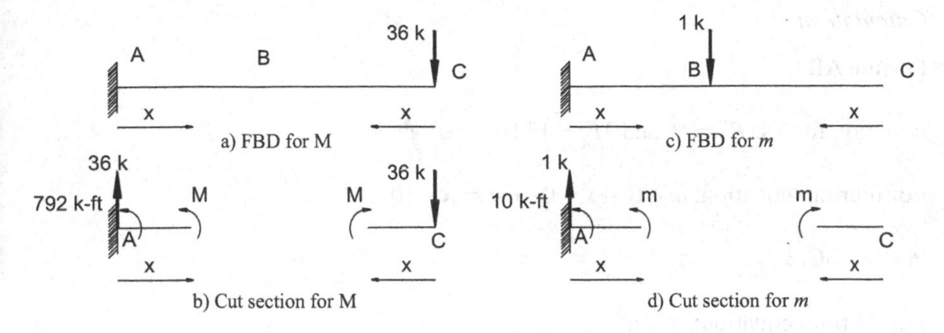

FIGURE 6.4 FBD for M and m (Δ_B) and cut sections.

The values for M and m will depend on the origin chosen and the corresponding change in limits. In doing this, it may simplify the integration and the final value of deflection will be the same.

Step 4

Deflection at B due to the applied loads

The required values for calculation of deflection are given in Table 6.1.

TABLE 6.1
Calculation of deflection $(\Delta_B)_L$ using Figures 6.3a and 6.3b

Portion of the beam	AB	BC
Origin	A	C
Limit	$x = 0\, to\, x = 10$	$x = 0\, to\, x = 12$
M under applied load	$36x - 792$	$-36x$
m *under unit load*	$x - 10$	0

Calculate M

Portion AB:

Reaction forces: $R_A = 36\ \boldsymbol{k}$ and

$$M_A = 36*(10+12) = 792^{k-ft}$$

Equilibrium equation:

$$M + 792 - 36x = 0 \rightarrow \boldsymbol{M} = 36\boldsymbol{x} - 792$$

Portion BC:

Equilibrium equation: $-M - 36x = 0 \rightarrow \boldsymbol{M} = -36\boldsymbol{x}$

Calculate m

Portion AB:

Reaction forces: $R_A = 1k$ and $M_A = 1*10 = 10^{k-ft}$

Equilibrium equation: $m + 10 - x = 0 \rightarrow \boldsymbol{m = x - 10}$

Portion BC:

Equilibrium equation: $\boldsymbol{m} = 0$

It is to be noted that the value of M (shown in Table 6.1) is calculated using Figure 6.4a and 6.4b, while the value of m is calculated from Figure 6.4c and 6.4d. The determinate structure shown in Figure 6.4c is the same determinate structure as shown in Figure 6.4a but acted on by a unit downward load at B (with no given applied loads) as it is assumed that the vertical deflection at B is downward. If at the end of the calculation the deflection at B comes out to be positive, it means that the actual deflection is downward. On the other hand, if the final deflection at B comes out to be negative, it means that the actual deflection at B is upward.

Substituting the values of M and m (from Table 6.1) in equation 6.3, the deflection $(\Delta_B)_L$ (deflection at B due to the applied loads) is calculated as,

$$\Delta_B = \Sigma \int \mathrm{Mm} \frac{dx}{EI} = \frac{1}{EI} \int_0^{10} (36x - 792)(x - 10)\,dx \tag{6.6}$$

$$\Delta_B = \frac{33600}{EI} \tag{6.7}$$

The *deflection at B due to a unit value of the redundant force* R_B (δ_{bb}) is obtained from Figure 6.5 as shown in Table 6.2.

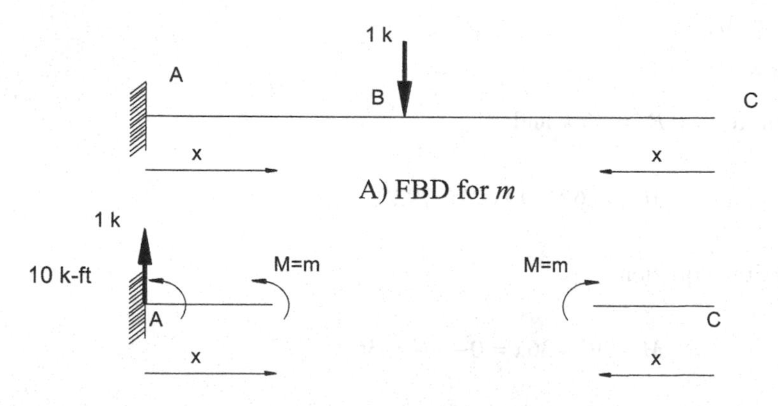

FIGURE 6.5 FBD for m (δ_{bb}) and cut section.

TABLE 6.2
Calculation of deflection (δ_{bb}) using Figure 6.4

Portion of the beam	AB	BC
Origin	A	C
Limit	$x = 0$ to $x = 10$	$x = 0$ to $x = 12$
$M = m$	$x - 10$	0

Substituting the values of M and m ($M = m$) in equation (1.3), we have

$$\delta_{bb} = \Sigma \int \text{Mm} \frac{dx}{EI} = \frac{1}{EI} \int_0^{10} (x-10)(x-10)\,dx \tag{6.8}$$

The deflection at B due to a unit value of the redundant force (R_B) is obtained as,

$$\delta_{bb} = \frac{1000}{3EI} \tag{6.9}$$

Step 5

Equation of compatibility of displacement at joint B requires that,

$$(\Delta_B)_L + (\Delta_B)_R = 0 \tag{6.10}$$

where

$$(\Delta_B)_R = R_B * (\delta_{bb})$$

This equation essentially says that the total vertical displacement at B has to be zero as it is a roller joint.

Substituting the values of $(\Delta_B)_L$ and (δ_{bb}) calculated above, equation (6.10) can be rewritten as,

$$\frac{33600}{EI} + R_B \times \frac{1000}{3EI} = 0 \tag{6.11}$$

Step 6

Solving equation 6.11 above, R_B can be obtained as negative which means,

$$R_B = 100.8\boldsymbol{k} \uparrow$$

This shows that R_B is upward, not downward, as assumed in Figure 6. 3c.

Step 7

Once the redundant force (R_B) is obtained, then the remaining reactions at A (R_A and M_A) can easily be obtained from equilibrium equations.

They are calculated using Figure 6.6 as,

$$\overset{+}{\rightarrow} \Sigma F_x = 0 \rightarrow A_x = 0$$

$$+\uparrow \Sigma F_y = 0 \rightarrow A_y + 100.8 - 36 = 0 \rightarrow A_y = -64.8\ \boldsymbol{k} \downarrow$$

$$\overset{+}{\circlearrowleft} \Sigma M_A = 0 \rightarrow M_A + 100.8(10) - 36(22) = 0 \rightarrow \boldsymbol{M_A} = -216^{k\text{-}ft} \circlearrowleft$$

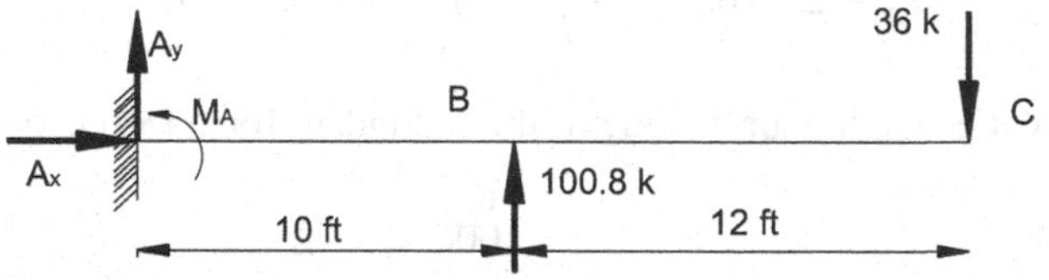

Ay
MA
B
36 k
C
Ax
100.8 k
10 ft
12 ft

FIGURE 6.6 Final reactions for the indeterminate beam.

It has been shown above by solving a simple example that when solving a statically indeterminate structure by the force method, first, write the correct expressions for M and m, and then integrate the expression to solve for deflection within the specified limits (consistent with the chosen origin).

6.5.3 Structures with Several Redundant Forces

As stated earlier, it is to be noted that if a structure has several redundant forces (i.e., $SI \geq 1$), then indeterminate structural analysis of the structure would involve obtaining redundant forces through solution of simultaneous equations. This will be followed by obtaining the remaining reactions at the supports (other than the redundant forces) through principles of statics as done in example 6.5.2.1.

The reader is advised to see other literature for detailed information such as those found in the references of this book.

6.6 APPLICATION OF THE FORCE METHOD TO INDETERMINATE FRAMES

The basic procedure for analysis of statically indeterminate frames essentially remains the same as outlined in Section 6.5, and as illustrated for a beam in Example 6.5.2.1.

Although the analysis of an indeterminate frame is, conceptually, very much similar to that of the beam, a frame consists of beams and columns so the analysis is slightly more complicated. After following the example below, it will be clear how to apply the force method to indeterminate frames.

6.6.1 EXAMPLES OF AN INDETERMINATE FRAME

A structural analysis dealing with a statically indeterminate frame by the force method is shown in Example 6.6.1.1.

Example 6.6.1.1

Determine the reactions at the supports of the frame shown in Figure 6.7 using the force method. A = 100 in^2, E = 29000 ksi and I = 833 in^4.

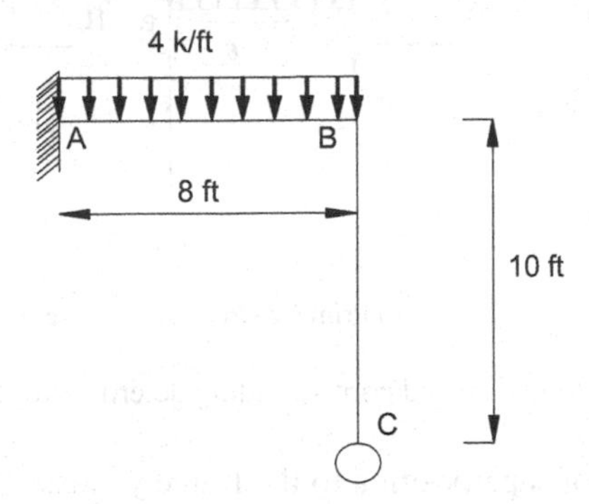

FIGURE 6.7 Statically indeterminate frame.

Solution

The procedure followed is as stated in Section 6.5.

Step 1

$$SI = n_u - n_e = 4 - 3 = 1$$

Step 2

Choose M_A as the redundant moment.

Step 3

The given statically indeterminate structure is split into two determinate structures as shown in Figure 6.8, with the redundant moment removed and with the applied loading as shown in Figure 6.8c. In Figure 6.8c, the frame is acted on by the redundant moment M_A. The rotational deflection at A due to applied loads is θ_A, and due to the unit load, M_A, is $M_A \times \alpha_{AA}$. These rotations can be obtained from equation (6.3) by finding the values of M using Figure 6.8b and respective values of m using Figure 6.8c. In both cases, H_A and V_A are found using static equilibrium equations.

Note on symbols: In general, (δ_{HAHA}) represents the horizontal deflection at A due to horizontal unit load at A (i.e., $H_A = 1$). Similarly, (δ_{VAHA}) represents the vertical deflection at A due to horizontal unit load at A. Along the same lines, (δ_{HAVA}) and (δ_{VAVA}) represent the horizontal and vertical deflection at A, respectively, due to a unit vertical load at A (i.e., $V_A = 1$).

Note: The values for M and m will depend on the origin chosen (with the corresponding change in limits). As can be expected, the final value of deflection will be the same irrespective of how it is done.

This frame is statically indeterminate to the first degree. Since we chose the moment reaction at A as the redundant, the support at A will become a pin as shown in Figure 6.8.

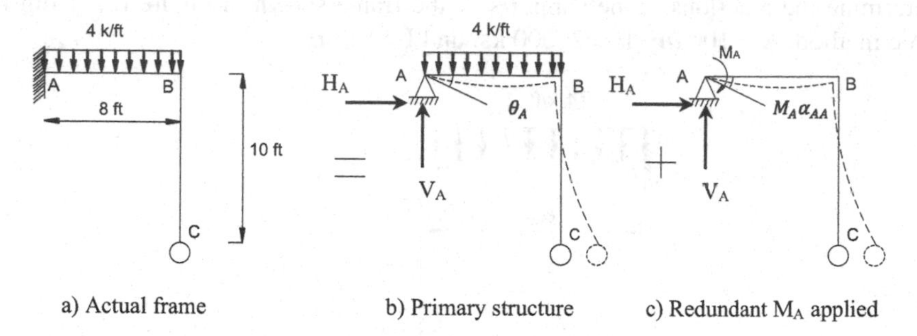

FIGURE 6.8 Given indeterminate and corresponding determinate structures.

Applying the principle of superposition to the frame yields:

$$-\theta_A - M_A * \alpha_{AA} = 0 \tag{6.12a}$$

In this case, both θ_A and $M_A * \alpha_{AA}$ are negative because they both create a clockwise rotation at joint A. This is negative by the sign convention defined in Section 6.5.1. Equation (6.12a) can also be written as:

$$\theta_A + M_A * \alpha_{AA} = 0 \tag{6.12b}$$

Equation (6.12b) can also be found by considering that both θ_A and $M_A * \alpha_{AA}$ create compression at the top fiber of member AB.

Step 4

Use the unit load method to calculate $\boldsymbol{\theta}_A$ (Figure 6.9 and Table 6.3):

TABLE 6.3
Deflection calculation for θ_A

Portion of the beam	AB	BC
Origin	A	C
Limit	$x = 0$ to $x = 8$	$x = 0$ to $x = 10$
M	$16x - 2x^2$	0
m_θ	$1 - \frac{x}{8}$	0

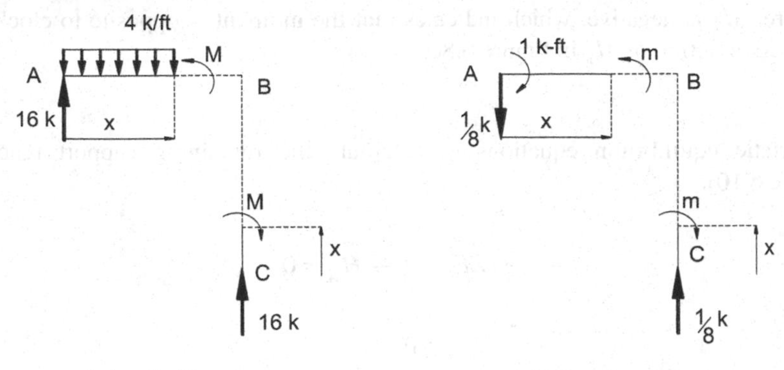

a) Applied loads b) Unit load M_A

FIGURE 6.9 Bending moments due to applied and unit loads.

$$\rightarrow \theta_A = \Sigma \int \text{Mm} \frac{dx}{EI} = \frac{1}{EI} = \int_0^8 (16x - 2x^2)\left(1 - \frac{x}{8}\right)dx = \frac{256}{3EI}$$

Use virtual work (unit load method) to calculate $\boldsymbol{\alpha}_{AA}$ (Table 6.4):

TABLE 6.4
Deflection calculation for α_{AA}

Portion of the beam	**AB**	**BC**
Origin	A	C
Limit	$x = 0$ to $x = 8$	$x = 0$ to $x = 10$
$M = m_\theta$	$1 - \frac{x}{8}$	0

$$\rightarrow \alpha_{AA} = \Sigma \int \text{m}_\theta \text{m}_\theta \frac{dx}{EI} = \frac{1}{EI} = \int_0^8 \left(1 - \frac{x}{8}\right)^2 dx = \frac{8}{3EI}$$

Step 5

Equation of compatibility:

$$\theta_A + M_A * \alpha_{AA} = 0$$

Step 6

Plugging in the values for deflection:

$$\frac{256}{3EI} + M_A\left(\frac{8}{3EI}\right) = 0 \rightarrow \boldsymbol{M}_A = -32^{k\text{-}ft} \circlearrowleft$$

Here, M_A is negative, which indicates that the moment is opposite to clockwise assumed direction of M_A in Figure 6.8c.

Step 7

Use static equilibrium equations to calculate the remaining support reactions (Figure 6.10):

$$\overset{+}{\rightarrow}\Sigma F_x = 0 \rightarrow \boldsymbol{H_A} = 0$$

$$\overset{+}{\circlearrowleft}\Sigma M_A = 0$$

$$M_A - 32(4) + V_C(8) = 0 \rightarrow \boldsymbol{V_C} = 12\,\boldsymbol{k} \uparrow$$

$$+\uparrow \Sigma F_y = 0:$$

$$V_A + 12 - 32 = 0 \rightarrow \boldsymbol{V_A} = 20\ \boldsymbol{k} \uparrow$$

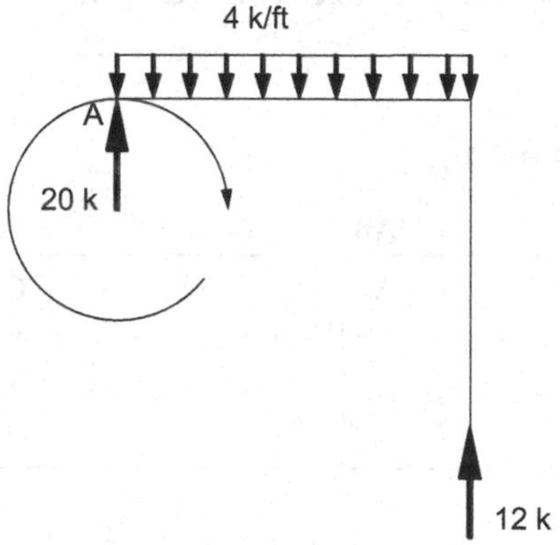

FIGURE 6.10 Final reactions and moments for the indeterminate frame.

This completes the solution of the problem.

6.7 APPLICATION OF FORCE METHOD TO ANALYSIS OF INDETERMINATE TRUSSES

The analysis procedure for a statically indeterminate truss follows the same lines of beams and frames discussed in Section 6.3. The basic equation used for calculating deflection is given by equation (6.4) and stated here again as,

$$\Delta = \Sigma \frac{NnL}{AE} \tag{6.4}$$

An example dealing with the analysis of a statically indeterminate truss is solved in Example 6.7.1.

EXAMPLE 6.7.1

Determine the reactions at the supports of the truss shown in Figure 6.11 using the force method. AE is constant.

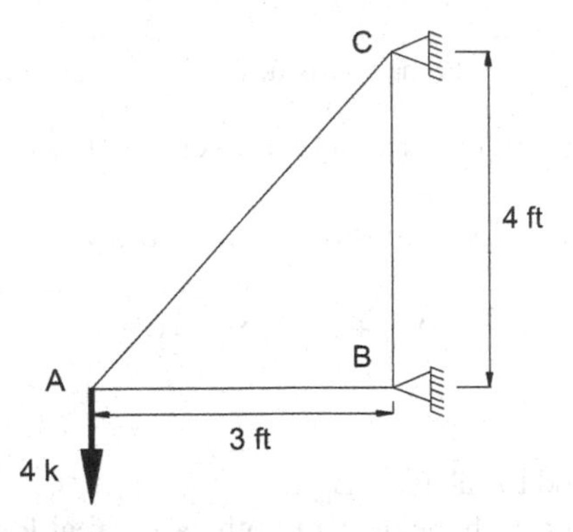

FIGURE 6.11 Statically indeterminate truss.

Solution

Step 1

Degree of indeterminacy $= b + r - 2j$

$$= 3 + 4 - 2(3) = 1$$

Step 2

Choosing BC as the redundant, this member will be "cut" to make the truss statically determinate.

Step 3

The given statically indeterminate structure is split into two determinate structures as shown in Figure 6.12. Figure 6.12b shows the structure under the given loading, and Figure 6.12c shows the truss with the redundant unit load applied.

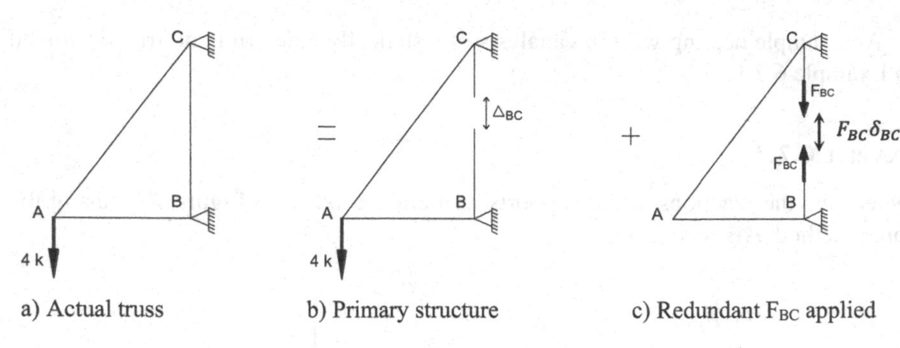

a) Actual truss b) Primary structure c) Redundant F_{BC} applied

FIGURE 6.12 Statically indeterminate and corresponding determinate trusses.

Applying the principle of superposition to the truss yields:

$$\Delta_{BC} + F_{BC} * \delta_{BC} = 0 \tag{6.13}$$

Step 4

Use the unit load method to calculate Δ_{BC}:

Calculate N and n for each member in both cases: real load and virtual unit load as shown in Figure 6.13.

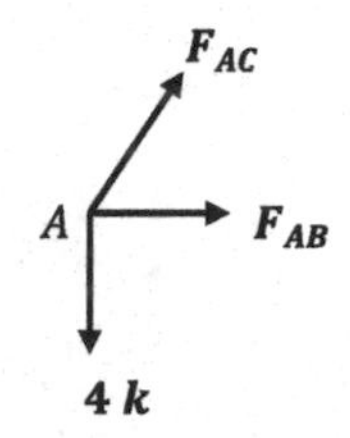

a) Calculation of F_{AB} and F_{AC}

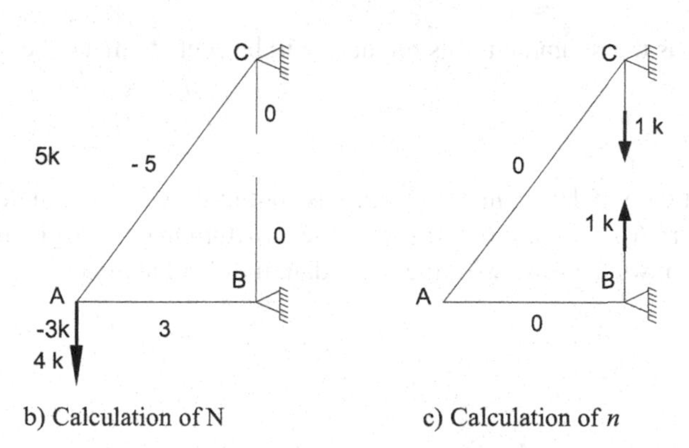

b) Calculation of N c) Calculation of n

FIGURE 6.13 Calculation of forces for N and n.

$$\Delta_{BC} = \Sigma \frac{nNL}{AE} + \frac{0(5)5}{AE} + \frac{0(-3)3}{AE} + \frac{1(0)4}{AE} = 0 \tag{6.14}$$

Use the unit load method to calculate $\boldsymbol{\delta}_{BC}$:

$$\delta_{BC} = \Sigma \frac{n^2 L}{AE} = \frac{0^2(5)}{AE} + \frac{0^2(3)}{AE} + \frac{1^2(4)}{AE} = \frac{4}{AE} \tag{6.15}$$

Step 5

The compatibility equation given by equation (6.13) is repeated below.

$$\Delta_{BC} + F_{BC} * \delta_{BC} = 0 \tag{6.13}$$

Step 6

From equation (6.13) $\rightarrow 0 + F_{BC} * \frac{4}{AE} = 0 \rightarrow \boldsymbol{F}_{BC} = 0$

Using this result, forces in other members and the support reactions can be calculated easily using the method of joints.

Step 7

The method of joints for B and C along with the final reactions is shown in Figure 6.14.

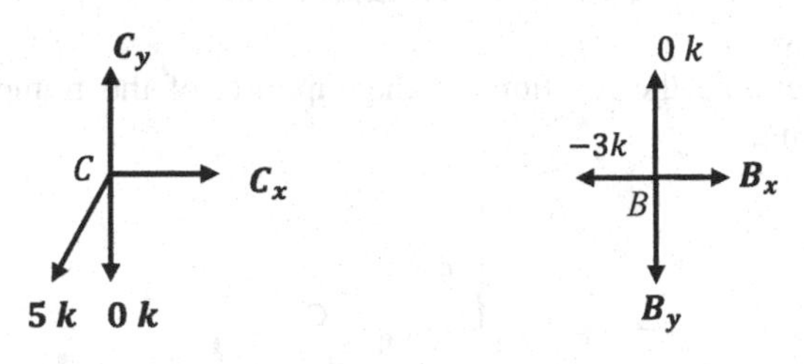

a) Calculation of C_x and C_y b) Calculation of B_x

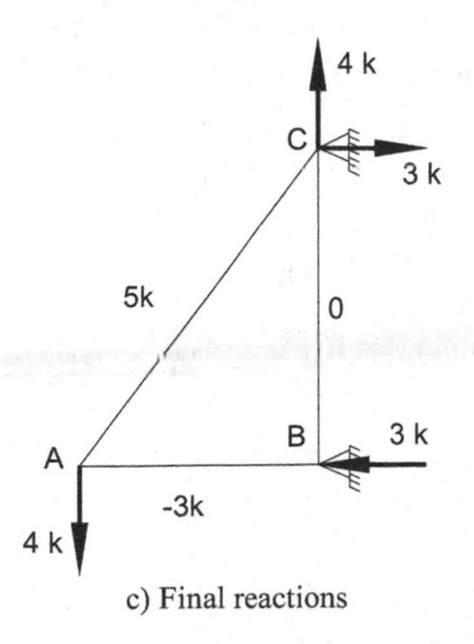

c) Final reactions

FIGURE 6.14 Final reactions and internal forces for the indeterminate truss.

6.8 SUMMARY

In this chapter, the basic concept of the force method is explained briefly but succinctly. This is followed by application of the force method to a set of problems dealing with structural analysis of an indeterminate beam, frame, and truss. It is to be noted that force method uses the concept of SI and involves a large number of deflection calculations. Hence, the knowledge of the prerequisite courses dealing with deflection calculations is paramount to a strong understanding of this approach.

PROBLEMS FOR CHAPTER 6

Analyze the problems from 6.1 to 6.3 for all the unknown reactions using the force method.

Problem 6.1 Determine the reactions at the supports of the beam shown in this figure. EI is constant.

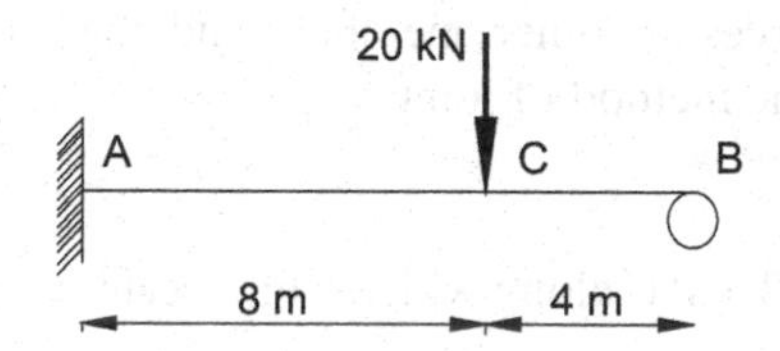

Problem 6.2 Determine the reactions at the supports of the frame shown in this figure. EI is constant.

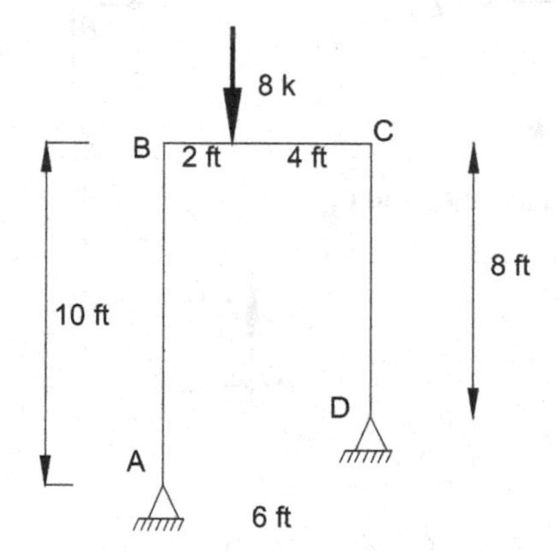

Problem 6.3 Determine the reactions at the supports of the truss shown in this figure. AE is constant.

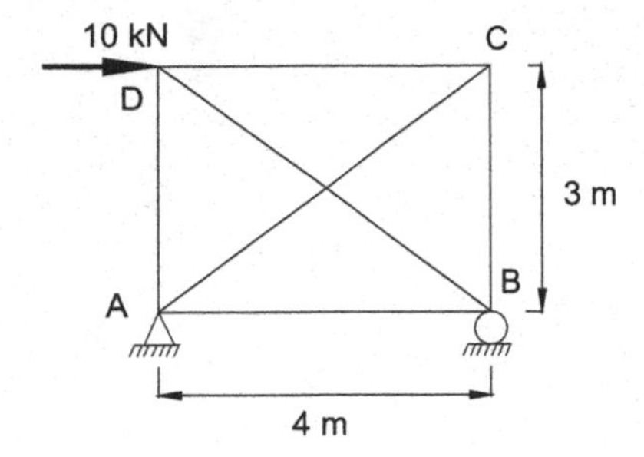

7 Displacement Method of Analysis
Slope-Deflection Method

7.1 BASIC CONCEPTS OF THE DISPLACEMENT METHOD

The displacement method refers to the general approach of solving indeterminate structural analysis problems with displacements as the primary variables. Two displacement methods that will be explained in this book are classical methods called slope-deflection and moment distribution. The displacement method uses the concept of structural kinematic indeterminacy (KI). The formula for this is as follows:

$$\text{KI} = \Sigma \text{ (degrees of freedom at all supports in the given structure)} \qquad (7.1)$$

where degrees of freedom are unrestrained motions of a joint/support. This means a fixed support has zero degrees of freedom and a pin has one (rotation).

The results obtained using the slope-deflection method are the end moments (internal moments) at the supports of the structure. These are found through a two-step process of, first, finding the rotations (slopes) and, second, finding the end moments. In contrast, the moment distribution method, which will be discussed in Chapter 8, gives end moments directly as a result of the procedure. After finding the end moments, the reactions at various supports can be determined using the principles of statics.

In the slope-deflection method, the unknown displacements are usually rotational displacements of a pin or roller support. The displacements are written in terms of the loads using the load-displacement relationships, also known as slope-deflection equations. The resulting equations are then solved for the displacements. Therefore, the main intermediate output resulting from the slope-deflection method is displacements. The final output is end moments.

7.2 BASIC PROCEDURE OF THE SLOPE-DEFLECTION METHOD

7.2.1 Slope-Deflection Equations

Before the actual procedure is discussed, it is important to introduce the slope-deflection equations, which are key to the slope-deflection method. Derivation of the slope-deflection equations will not be shown here; these are done with great detail in the books listed in the references section.

DOI: 10.1201/9781003246633-9

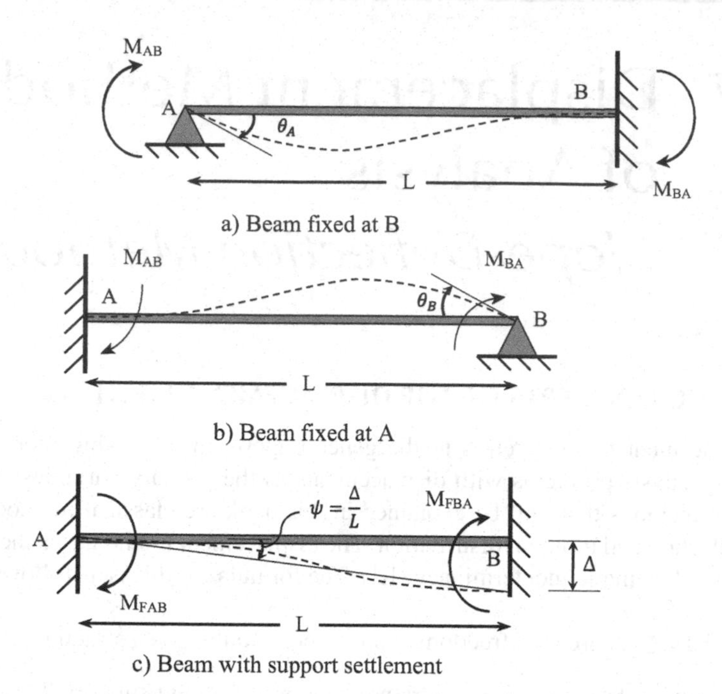

FIGURE 7.1 Moments and displacements on typical indeterminate beams.

With respect to Figure 7.1, the slope-deflection equations can be written (without support settlement) as,

$$M_{AB} = M_{FAB} + \frac{2EI}{L}\left(2\theta_A + \theta_B\right) \tag{7.2}$$

$$M_{BA} = M_{FBA} + \frac{2EI}{L}\left(2\theta_B + \theta_A\right) \tag{7.3}$$

Where M_{FAB} and M_{FBA} are assumed to be acting clockwise.

Wang (1953) advocates using relative stiffness factors instead of actual stiffness factors to simplify the calculations. Modifying equations (7.2) and (7.3) to include stiffness factors yields,

$$M_{AB} = M_{FAB} + K_{AB}\left(2\theta_A + \theta_B\right) \tag{7.2a}$$

$$M_{BA} = M_{FBA} + K_{BA}\left(2\theta_B + \theta_A\right) \tag{7.3a}$$

where M_{AB} = Moment at joint A of member AB
M_{BA} = Moment at joint B of member AB
M_{FAB} = Fixed-end moment at end A of member AB due to applied loading
M_{FBA} = Fixed-end moment at end B of member AB due to applied loading
θ_A = Slope at joint A
θ_B = Slope at joint B.

7.2.2 Sign Convention for Displacement Methods

- Clockwise moments are positive.
- Counterclockwise moments are negative.
- Clockwise rotations are positive.
- Counterclockwise rotations are negative.

7.2.3 Fixed-End Moments

Fixed-end moments are moment reactions of a single span beam having fixed supports for a given loading. Table 1A give fixed-end moment values for various load types.

7.3 ANALYSIS OF CONTINUOUS BEAMS BY THE SLOPE-DEFLECTION METHOD

Before discussing examples, the calculation procedure will be outlined below:

1. Calculate all the fixed-end moments due to applied loads at the end of each span using Table 1A found in the appendix.
2. Calculate the KI of the structure. It is expressed as K.I. = Σ (degrees of freedom at all supports in the given structure).

   Degrees of freedom are unrestrained motions of a joint/support. This means a fixed support has zero degrees of freedom, a pin has one (rotation), and a frame's joint has one (rotation).
3. Formulate all the slope-deflection equations for each member of the continuous beam using equations (7.2) and (7.3). These equations are in terms of the unknown rotations at the supports.
4. Formulate simultaneous equilibrium equations at the joints (not fixed) using the basic premise that the sum of the end moments at the support (for all the members joining at the support) is zero. The number of unknown rotations in the problem is equal to the number of simultaneous equations to be solved as well as the KI found in step 2.
5. Solve the simultaneous equations formulated in step 4 and obtain rotations at the supports.
6. Compute end moments by substituting rotations back into the slope-deflection equations.
7. Depending on the statement of the problem, calculate all the reactions.
8. Draw shear and moment diagrams for the continuous beam as needed.

Example 7.3.1

Determine the reactions at the supports for the statically indeterminate beam shown in Figure 7.2 by the slope-deflection method. Take $E = 29000$ ksi and $I = 446$ in^4.

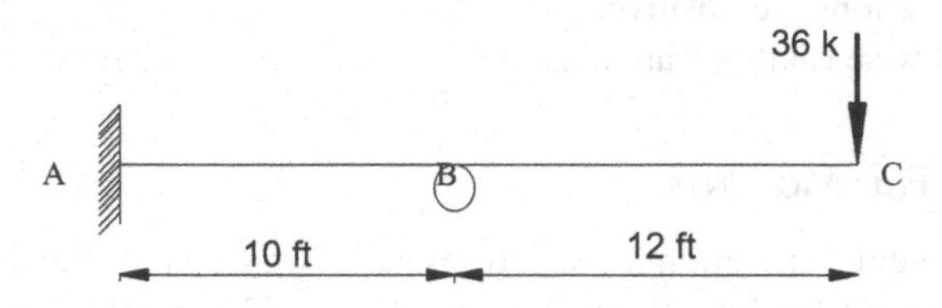

FIGURE 7.2 Statically indeterminate beam.

Solution

Step 1

Calculate the fixed-end moments using Table 1A found in the appendix. The fixed-end moments for AB and BA are both zero because there is no loading on the span of member AB.

$$M_{FAB} = 0 \text{ and } M_{FBA} = 0$$

Step 2

$$KI = 1$$

The unknown displacement is θ_B. Although the other unknown displacements (θ_C and Δ_C) exist, these displacements are unnecessary to solve the problem because they do not occur at a support where specific unknowns need to be found (B_y). θ_B must be found so that we can find the reaction at B. In contrast, solving for θ_c would not give us any information about the reactions of the structure.

Step 3

Slope-deflection equations are formed using equations (7.2) and (7.3) as,

$$M_{AB} = 0 + 2E\frac{I}{10}\left[2\theta_A + \theta_B\right] = \frac{EI}{5}\theta_B \tag{7.4}$$

(Note: $\theta_A = 0$ due to the fixed support at A)

$$M_{BA} = 0 + 2E\frac{I}{10}\left[2\theta_B + \theta_A\right] = \frac{2EI}{5}\theta_B \tag{7.5}$$

Similarly, M_{BC} can be written as,

$$M_{BC} = -36 * 12 = -432^{k-ft}$$

Note: M_{BC} is negative because the internal moment caused by the loading acts in the counterclockwise direction (opposite to the external moment at that point).

Step 4

Since $KI = 1$ for this problem, there is only one unknown, which is θ_B. Hence, there is only one joint equilibrium equation to be solved. This is given as,

$$M_{BA} + M_{BC} = 0 \tag{7.6}$$

Step 5

Substituting the expressions for M_{BA} and M_{BC} from step 3, we have,

$$\frac{2EI}{5}\theta_B - 432 = 0 \rightarrow \theta_B = \frac{1080}{EI} \tag{7.7}$$

Step 6

Substituting the value of the rotation back into the slope-deflection equations found in step 3, the end moments can be expressed as,

$$M_{AB} = 216^{k\text{-}ft} \circlearrowright \tag{7.8}$$

$$M_{BA} = 432^{k\text{-}ft} \circlearrowright \tag{7.9}$$

Step 7

The reactions at A (A_y) and at B (B_y) are calculated from principles of statics as shown in Figure 7.3:

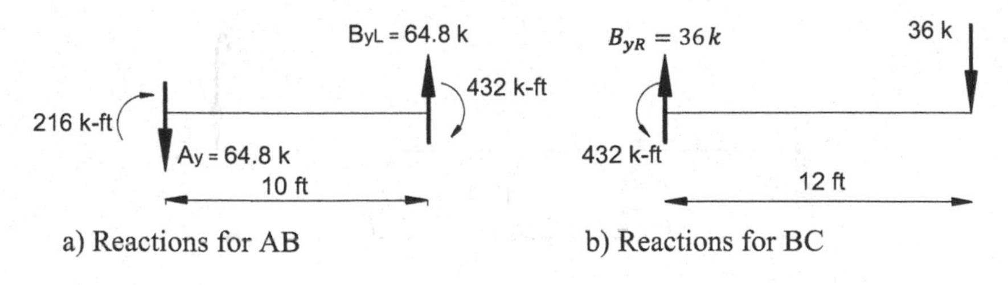

FIGURE 7.3 Reaction calculation.

$$\overset{+}{\rightarrow}\Sigma F_x = 0 \quad A_x = 0$$

$$\overset{+}{\circlearrowright}\Sigma M_B = 0 : \text{Member AB}$$

$$216 + 432 - A_y * 10 = 0$$

$$A_y = \frac{216 + 432}{10} = 64.8\,\boldsymbol{k} \downarrow$$

$$+\uparrow \Sigma F_y = 0 : \text{Whole beam}$$

$$-A_y + B_y - 36 = 0$$

$$-64.8 + B_y - 36 = 0$$

$$B_y = 100.8\,\boldsymbol{k} \uparrow$$

Step 8

The bending moment diagram (BMD) can be drawn as needed.

7.4 ANALYSIS OF CONTINUOUS BEAMS WITH SUPPORT SETTLEMENTS BY THE SLOPE-DEFLECTION METHOD

The slope-deflection equations including settlement with respect to Figure 7.4 are given as,

$$M_{AB} = M_{FAB} + \frac{2EI}{L}\left(2\theta_A + \theta_B - 3\Psi_{AB}\right) \tag{7.10}$$

$$M_{BA} = M_{FBA} + \frac{2EI}{L}\left(2\theta_B + \theta_A - 3\Psi_{BA}\right) \tag{7.11}$$

where M_{AB} = Moment at joint A of member AB
M_{BA} = Moment at joint B of member AB
M_{FAB} = Fixed-end moments at the end A of member AB due to applied loading
M_{FAB} = Fixed-end moments at the end B of member AB due to applied loading
θ_A = Slope at joint A
θ_B = Slope at joint B
Ψ_{AB} = Rotation of the member AB due to translation (settlement) of joint B perpendicular to member AB

$$\psi_{\mathrm{AB}} = \Delta / L \tag{7.12}$$

where Δ = Translation (settlement) of joint B perpendicular to axis of member AB
L = Length of member AB

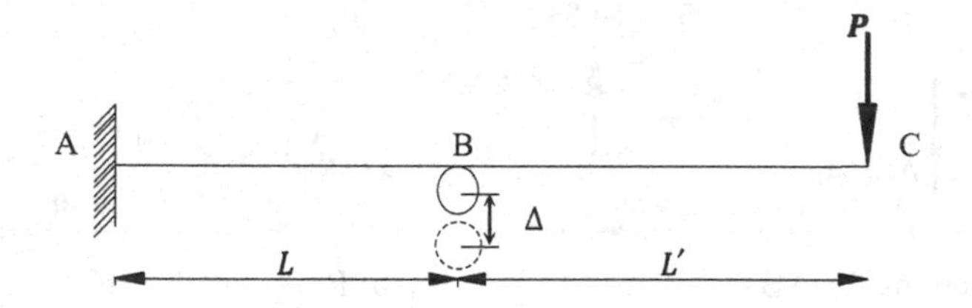

FIGURE 7.4 Statically indeterminate beam with support settlement at B.

It is to be noted that Ψ is treated positive when the rotation is clockwise, consistent with the sign convention stated in Section 7.2.2.

Equations (7.10) and (7.11) can be rewritten using the relative stiffness factors and relative Ψ_{AB} values as,

$$M_{AB} = M_{FAB} + K_{AB}\left(2\theta_A + \theta_B - 3\Psi_{rel}\right) \tag{7.10a}$$

$$M_{BA} = M_{FBA} + K_{BA}\left(2\theta_B + \theta_A - 3\Psi_{rel}\right) \tag{7.11a}$$

The relative stiffness factors (K_{AB} and K_{BA}) for any general member AB can be expressed as $2EI/L$ and Ψ_{rel} as Δ/L.

The procedure for solving continuous beams where joints are subjected to vertical translation amounting to settlement of supports remains the same as discussed in Section 7.3.

Example 7.4.1

Determine the reactions at the supports for the statically indeterminate beam shown in Figure 7.4 by the slope-deflection method. Take $E = 29000$ ksi and I = 446 in^4. The support at B is displaced downward 1 in.

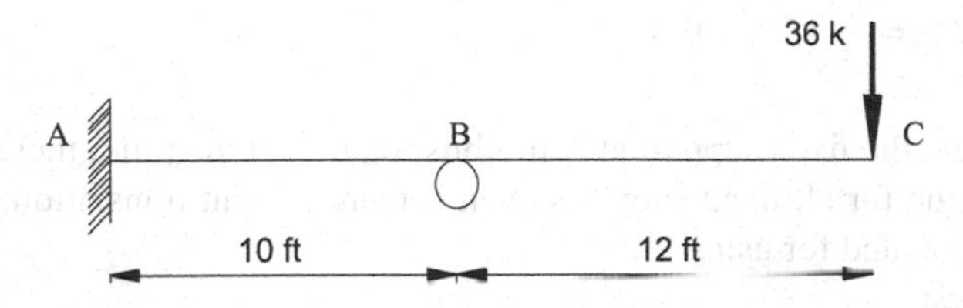

FIGURE 7.4a (repeated): Statically indeterminate beam with support settlement at B.

Solution

Step 1

In the slope-deflection method, *fixed end moments due to support settlement are not considered* because support settlement is accounted for using Ψ.

$$M_{FAB} = 0 \text{ and } M_{FBA} = 0$$

Step 2

$$KI = 1$$

Since Ψ is known, the only unknown displacement is θ_B. Moreover, due to downward displacement at B, it can be seen that the cord of span AB rotates clockwise; thus, Ψ is positive (Figure 7.5).

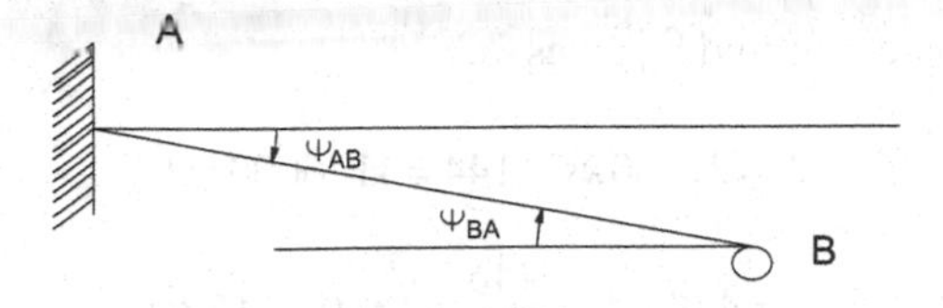

FIGURE 7.5 Effect of displacement at B.

$$\Psi_{AB} = \Psi_{BA} = \frac{1in}{10(12)in} = 0.00833\,rad$$

Step 3

Slope-deflection equations are formed using equations (7.10) and (7.11).

$$M_{AB} = 0 + 2E\frac{I}{10}\left[2\theta_A + \theta_B - 3\Psi_{AB}\right] = \frac{EI}{5}\left(\theta_B - 3*0.00833\right)$$

$$M_{AB} = \frac{EI}{5}\left(\theta_B - 0.025\right)$$

$$M_{BA} = 0 + 2E\frac{I}{10}\left[2\theta_B + \theta_A - 3\Psi_{BA}\right] = \frac{EI}{5}\left(2\theta_B - 3*0.00833\right)$$

$$M_{BA} = \frac{EI}{5}\left(2\theta_B - 0.025\right)$$

In this problem, the fixed support at A inhibits rotation at the joint; therefore, $\theta_A = 0$. This will be true for all fixed supports even if there is joint translation; the member rotation is accounted for using Ψ.

From statics:

$$M_{BC} = -36*12 = -432^{k-ft}$$

Note: M_{BC} is negative because the internal moment caused by the loading acts in the counterclockwise direction (opposite of the external moment at that point).

Step 4

The only joint equilibrium equation is for joint B and since it is a roller,

$$M_{BA} + M_{BC} = 0$$

Step 5

Substituting the expressions for M_{BA} and M_{BC} from step 3 we have

$$\rightarrow \frac{EI}{5}\left(2\theta_B - 0.025\right) - 432 = 0$$

Solving this equation to find θ_B gives

$$E = 29000\,ksi*144 = 4176000\,ksf$$

$$I = 446\,in^4 = \frac{446}{12^4}ft^4 = 0.0215\,ft^4$$

$$\rightarrow \theta_B = 0.02453\,rad$$

Step 6

Substituting the value of the rotation back into expressions for end moments calculated in step 3, the end moments can be expressed as

$$M_{AB} = -8.44^{k-ft} \circlearrowleft$$

$$M_{BA} = 432^{k-ft} \circlearrowright$$

Steps 7 and 8

The reactions at A (R_A) and B (R_B) are calculated using principles of statics as can be seen below (Figure 7.6).

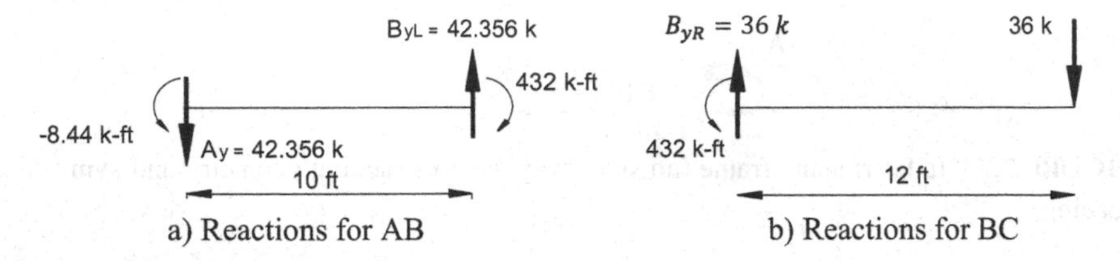

FIGURE 7.6 Reaction calculation.

$$\overset{+}{\rightarrow} \Sigma F_x = 0 \rightarrow A_x = 0$$

$$\overset{+}{\circlearrowright} \Sigma M_B = 0 : \text{Member AB}$$

$$-8.44 + 432 - A_y * 10 = 0$$

$$A_y = \frac{432 - 8.44}{10} = 42.356\, \boldsymbol{k} \downarrow$$

$$+\uparrow \Sigma F_y = 0 : \text{Whole beam}$$

$$-A_y + B_y - 36 = 0$$

$$-42.356 + B_y - 36 = 0$$

$$\boldsymbol{B_y} = 78.3560\, \boldsymbol{k} \uparrow$$

7.5 APPLICATION OF THE SLOPE-DEFLECTION METHOD TO ANALYSIS OF FRAMES WITHOUT JOINT MOVEMENT

The procedure for solving a statically indeterminate frame is the same as a statically indeterminate beam, which was explained in Section 7.3. An example is provided below to clarify the concept and procedure.

EXAMPLE 7.5.1

Determine the moments at each joint of the frame shown in Figure 7.7.

$E = 29000$ ksi, $A = 16\ in^2$, and $I = 446\ in^4$ for all members.

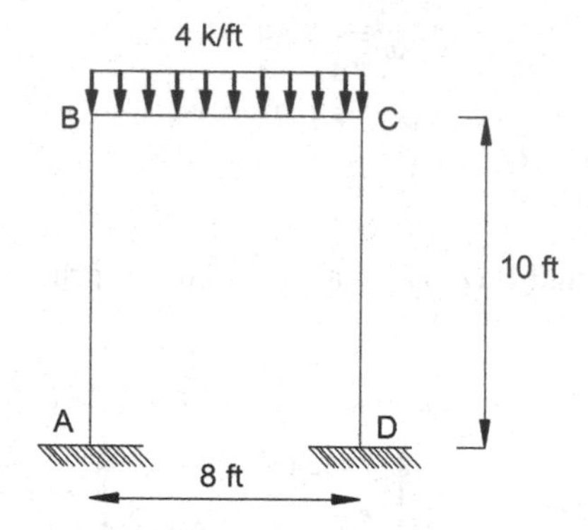

FIGURE 7.7 Indeterminate frame (no side sway due to structural geometry and symmetric loading).

Solution

Step 1

Since the loading is only on the span BC, there will only be fixed-end moments in members BC and CB.

$$M_{FBC} = -\frac{wL^2}{12} = -\frac{4(8)^2}{12} = -21.33^{k-ft}$$

$$M_{FCB} = \frac{wL^2}{12} = \frac{4(8)^2}{12} = 21.33^{k-ft}$$

Step 2

$$KI = 2$$

There are two unknown displacements in this problem, which are θ_B and θ_C. They are unknown because these frame joints will rotate as the members bend due to the applied loading. The rotations θ_A and θ_D are zero because of the fixed supports at A and D. Due to symmetrical loading, there will be no side sway in the frame, and therefore $\Psi = 0$.

Step 3

The slope-deflection equations are formulated below using equations (7.2) and (7.3) as

$$M_{AB} = 2E\frac{I}{10}\left[2\theta_A + \theta_B\right] = \frac{EI}{5}\theta_B$$

$$M_{BA} = 2E\frac{I}{10}\left[2\theta_B + \theta_A\right] \qquad = \frac{2EI}{5}\theta_B$$

$$M_{BC} = -21.33 + 2E\frac{I}{8}\left[2\theta_B + \theta_C\right] \;= -21.33 + \frac{EI}{4}\left(2\theta_B + \theta_C\right)$$

$$M_{CB} = 21.33 + 2E\frac{I}{8}\left[2\theta_C + \theta_B\right] \;= 21.33 + \frac{EI}{4}\left(\theta_B + 2\theta_C\right)$$

$$M_{CD} = 2E\frac{I}{10}\left[2\theta_C + \theta_D\right] \qquad = \frac{2EI}{5}\theta_C$$

$$M_{DC} = 2E\frac{I}{10}\left[2\theta_D + \theta_C\right] \qquad = \frac{EI}{5}\theta_C$$

Step 4

The corresponding joint equilibrium equations are written as

$$M_{BA} + M_{BC} = 0 \quad and \quad M_{CB} + M_{CD} = 0$$

Step 5

Substituting the expressions for M_{BA}, M_{BC}, M_{CB}, and M_{CD} from step 3, we have

$$\frac{2EI}{5}\theta_B - 21.33 + \frac{EI}{4}\left(2\theta_B + \theta_C\right) = 0 \tag{7.13}$$

$$21.33 + \frac{EI}{4}\left(\theta_B + 2\theta_C\right) + \frac{2EI}{5}\theta_C = 0 \tag{7.14}$$

Simplifying these equations to isolate θ_B and θ_C gives:

From (7.13) → $0.9EI\theta_B + 0.25EI\theta_C = 21.33$

From (7.14) → $0.25EI\theta_B + 0.9EI\theta_C = -21.33$

Solving equations (7.13) and (7.14) yields:

$$\theta_B = \frac{32.815}{EI} \quad and \quad \theta_C = -\frac{32.815}{EI}$$

This step of solving the simultaneous equations can be greatly simplified by using a calculator with this capability. Otherwise, hand calculations can be used, but these will not be shown in the text. Note that this problem can be simplified by recognizing that $\theta_C = -\theta_B$ from symmetry of the frame and only solving for one unknown.

Step 6

Substituting the value of the rotation back into expressions for end moments calculated in step 3, the end moments can be expressed as

$$M_{AB} = 6.563^{k-ft}$$

$$M_{BA} = 13.126^{k-ft}$$

$$M_{BC} = -13.126^{k-ft}$$

$$M_{CB} = 13.126^{k-ft}$$

$$M_{CD} = -13.126^{k-ft}$$

$$M_{DC} = -6.563^{k-ft}$$

Based on the moments of each joint, we can easily compute the reactions at support (Figure 7.8).

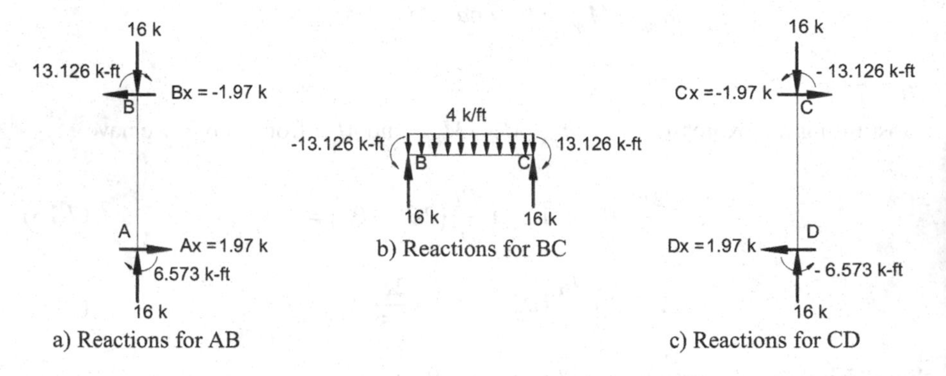

FIGURE 7.8 Reactions for the frame.

Alternatively, the relative stiffness factors could have been used to solve this problem. If one was to use this concept, the relative stiffness factors for AB and BC would be as follows:

$$K_{AB-rel} = \frac{2*I}{10}(20) = 4$$

$$K_{BC-rel} = \frac{2*I}{8}(20) = 5$$

Here, the relative stiffness factors $\left(\frac{2I}{L}\right)$ have been multiplied by the least common multiple (LCM) to simplify the calculations. Also, E and I are not included in the relative stiffness factors because they must be constant in all members to use K_{rel}.

The rotations obtained using this concept are different from those found using the actual stiffness factors because they are modified according to the LCM. The point to be noted is that the final end moments remain the same and calculation is facilitated.

If one was to use these relative stiffness factors and modified slope-deflection equations (7.2a) and (7.2b), the value of θ_B comes out to be 1.6408. However, the actual end moments remain the same.

$$M_{AB} = \frac{EI}{5}\theta_B = \frac{EI}{5} * \frac{32.815}{EI} = 6.563^{k-ft}$$

$$M_{AB-rel} = 4\left[\theta_B\right] = 4 * 1.6408 = 6.563^{k-ft}$$

7.6 DERIVATION OF SHEAR CONDITION FOR FRAMES (WITH JOINT MOVEMENT)

When analyzing frames with joint movement, an extra unknown (Δ or Ψ) is added to the usual unknown displacements. This means an extra equation is needed. The equation is obtained from what is known as the “shear condition” at the base supports of the frame.

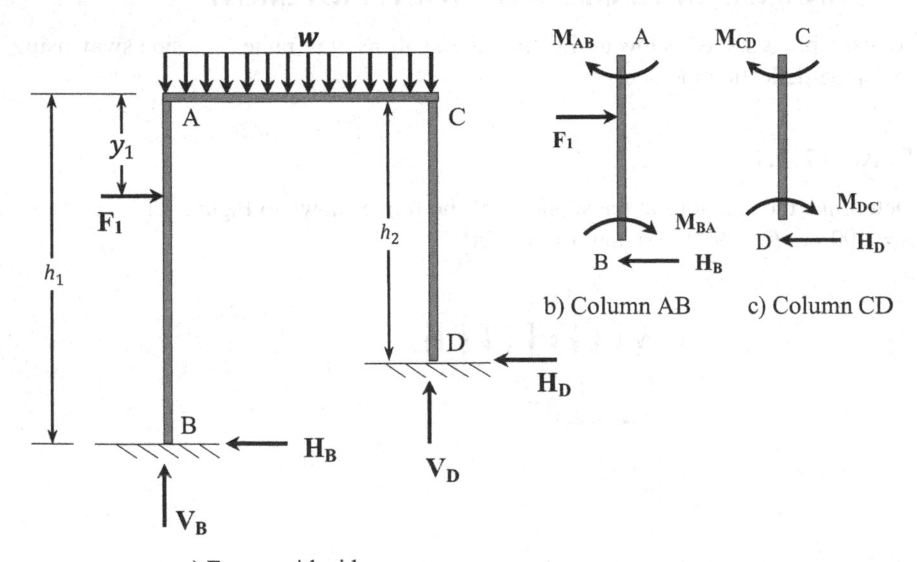

FIGURE 7.9 Frame with sidesway – Basic concept illustration.

For a typical frame (Figure 7.9), the shear condition obtained from the basic equation $\Sigma F_x = 0$ is given as

$$F_1 - H_B - H_D = 0 \tag{7.15}$$

where H_B and H_D can be found by taking $\sum M_A = 0$ and $\sum M_C = 0$ using Figures 7.9b and 7.9c.

This would yield the flowing two equations:

$$H_B = \frac{F_1 y_1}{h_1} - \frac{M_{AB} + M_{BA}}{h_1} \tag{7.15a}$$

$$H_D = -\frac{M_{CD} + M_{DC}}{h_2} \tag{7.15b}$$

Equations (7.15a) and (7.15b) are written with the assumption that the end moments of a column are clockwise (positive). Figure 7.9b and 7.9c show the free body diagrams for columns AB and CD with which the expression for H_B and H_D are derived. Equation (2.15) has to be solved in addition to the other joint equilibrium equations.

The rest of the procedure remains the same as outlined in Section 7.3.

7.7 APPLICATION OF THE SLOPE-DEFLECTION METHOD TO ANALYSIS OF FRAMES WITH JOINT MOVEMENT

An example is solved below to illustrate the analysis of a frame with side sway using the slope-deflection method.

EXAMPLE 7.7.1

Determine the reactions at the supports of the frame shown in Figure 7.10.
$A = 100$ in², $E = 29000$ ksi, and $I = 833$ in⁴.

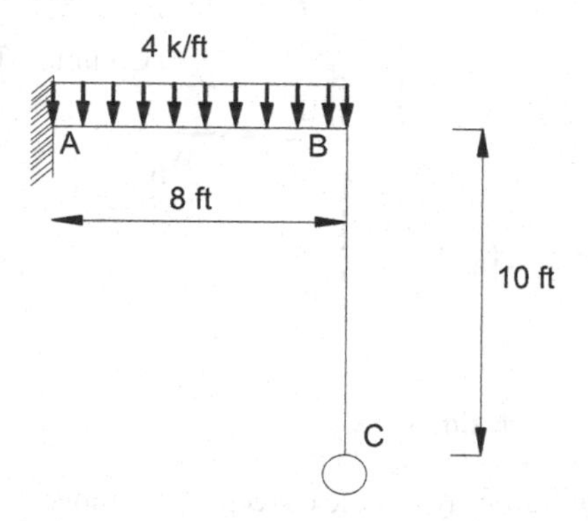

FIGURE 7.10 Indeterminate frame (sides sway).

Solution

Step 1

The fixed-end moments are calculated using Table 1A found in the appendix.

$$M_{FAB} = -\frac{4*8^2}{12} = -21.33^{k-ft}$$

$$M_{FBA} = \frac{4*8^2}{12} = 21.33^{k-ft}$$

Step 2

The unknown displacements are θ_B, θ_C, and Ψ_{BC} (Figure 7.11). $KI = 3$

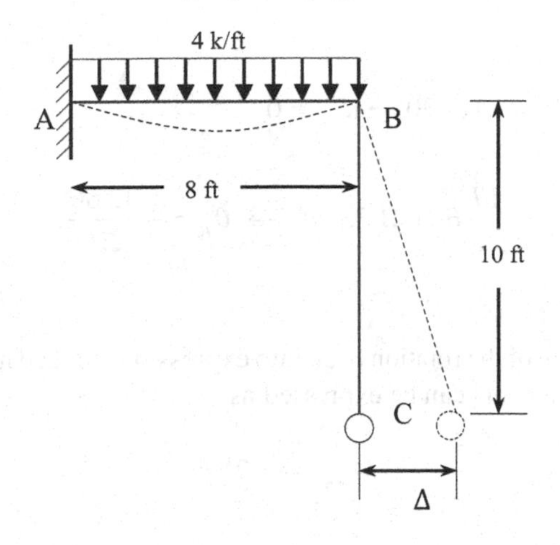

FIGURE 7.11 Bending of frame in Example 7.7.1.

Step 3

The slope-deflection equations for this structure can be written as

$$M_{AB} = 2E\frac{I}{8}\left[2(0)+\theta_B\right]-21.33 = \frac{EI}{4}\theta_B - 21.33 \tag{7.16}$$

(Note: $\theta_A = 0$ due to fixed support at A)

$$M_{BA} = 2E\frac{I}{8}\left[2\theta_B + 0\right]+21.33 = \frac{EI}{2}\theta_B + 21.33 \tag{7.17}$$

$$M_{BC} = 2E\frac{I}{10}\left[2\theta_B + \theta_C - \Psi_{BC}\right] \tag{7.18}$$

$$M_{CB} = 2E\frac{I}{10}\left[2\theta_C + \theta_B - \Psi_{BC}\right] \quad (7.19)$$

Step 4

$$\text{Moment equilibrium required : } M_{BA} + M_{BC} = 0 \quad (7.20)$$

$$\text{Roller support at C} \rightarrow M_{CB} = 0 \quad (7.21)$$

Due to symmetrical loading there is no moment in member BC, but the structure will still sidesway. Hence

$$M_{BC} = 0 \quad (7.22)$$

Step 5

From equations (7.18) and (7.20) $\rightarrow M_{BA} = 0$

$$\frac{EI}{2}\theta_B + 21.33 = 0 \rightarrow \boldsymbol{\theta_B} = -\frac{42.66}{\boldsymbol{EI}}$$

Step 6

Substituting the value of the rotation back into expressions for end moments calculated in step 3, the end moments can be expressed as

$$\boldsymbol{M_{AB}} = -32^{k\text{-}ft}$$

Steps 7 and 8

The reactions at A (A_x and A_y) and C (C_y) are calculated from principles of statics:

$$\overset{+}{\rightarrow}\Sigma F_x = 0 \rightarrow \boldsymbol{A_x} = 0$$

$$\Sigma M_A = 0:$$

$$-32 + 32*4 - Cy*8 = 0 \rightarrow \boldsymbol{C_y} = 12\boldsymbol{k}(\uparrow)$$

$$+\uparrow \Sigma F_y = 0:$$

$$A_y + 12 - 32 = 0 \rightarrow \boldsymbol{A_y} = 20\boldsymbol{k}(\uparrow)$$

After the reactions are obtained, the BMD and the shear force diagram can be drawn as needed.

7.8 SUMMARY

In this chapter, the fundamentals of a classical method, called the slope-deflection method, were discussed. This was followed by examples applying it to beams and frames. The slope-deflection method essentially consists of solving a set of simultaneous equations where the unknown values are displacements. Finally, end moments are calculated using these displacements. This method is easy to use and, unlike the force method, does not require knowing how to do deflection calculations.

PROBLEMS FOR CHAPTER 7

Analyze problems 7.1 to 7.3 using the slope-deflection method.

Problem 7.1 Determine the reactions at the supports of the beam shown in this figure. EI is constant.

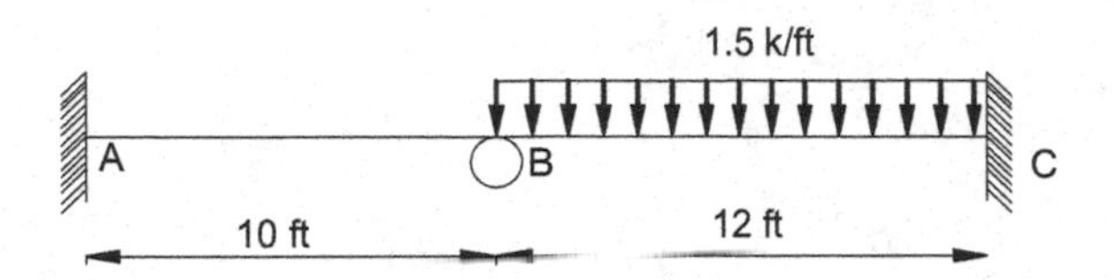

Problem 7.2 Determine the reactions at the supports of the frame shown in this figure. $A = 100$ in^2, $E = 29{,}000$ ksi, and $I = 833$ in^4.

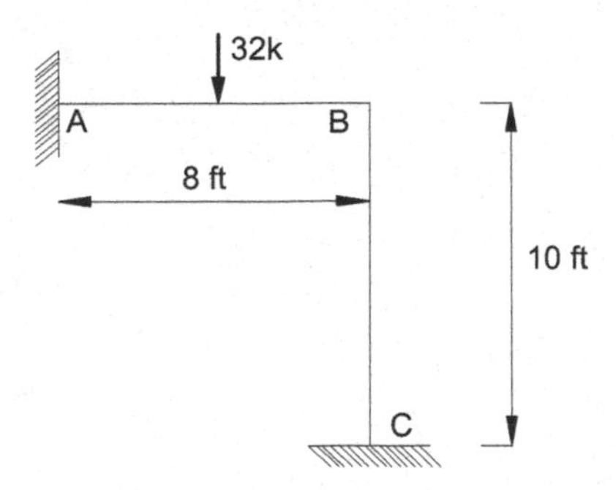

Problem 7.3 Determine the reactions at the supports of the frame shown in this figure. $A = 100\ \text{in}^2$, $E = 29{,}000$ ksi, and $I = 833\ \text{in}^4$

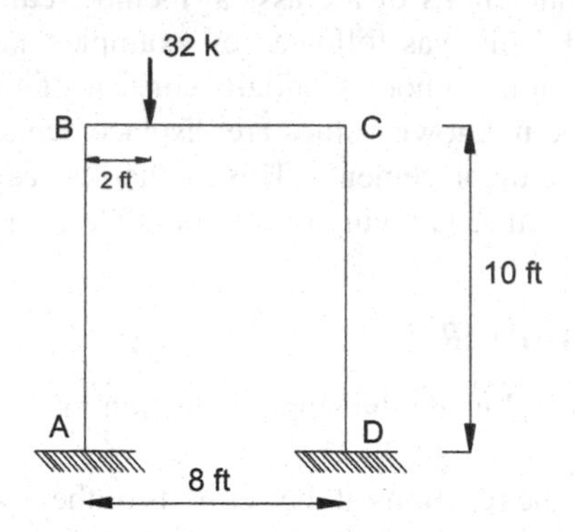

8 Displacement Method of Analysis
Moment Distribution Method

8.1 BASIC CONCEPTS OF MOMENT DISTRIBUTION METHOD

Hardy Cross originally developed the moment distribution method in 1930. It is a classical and iterative method. It essentially consists of locking and unlocking each joint consistent with the actual boundary conditions. This means that the whole procedure of moment distribution is carried out in such a way that at the end of it, the final end moments for a hinge (pin) joint should be zero while a fixed joint can have any amount of moment. Analysis of a structure essentially involves finding the end moments for each member. It will be interesting to compare the moment distribution method with another classical method called the slope-deflection method (discussed in Chapter 7). In the case of the slope-deflection method, finding end moments of members is a two-step process. The first step is finding the slopes at each joint and the second step is finding end moments for each member. On the other hand, the moment distribution method directly gives the end moments for each member. The moment distribution method, like the slope-deflection method, uses fixed-end moments and stiffness factors. Additionally, the moment distribution method uses distribution factors (DFs). It is through the DFs that the moment distribution is essentially carried out because they dictate how much moment a specific joint will transfer. Distribution factors are obtained using the stiffness factors for each member in such a way that it reflects the property of the joint. Thus, since the total moment at a hinge joint is zero, the distribution factor at a hinge joint is 1. Similarly, the distribution factor at a fixed joint is zero as the fixed joint can carry any amount of moment. The distribution factor will be discussed in more detail in Section 8.2.3.

8.2 STIFFNESS FACTOR, CARRY-OVER FACTOR, AND DISTRIBUTION FACTOR

Three important factors used in the moment distribution method are the stiffness factor (K), carry-over factor (CO), and the distribution factor (DF). These will be described in the following sections.

DOI: 10.1201/9781003246633-10

8.2.1 Stiffness Factor

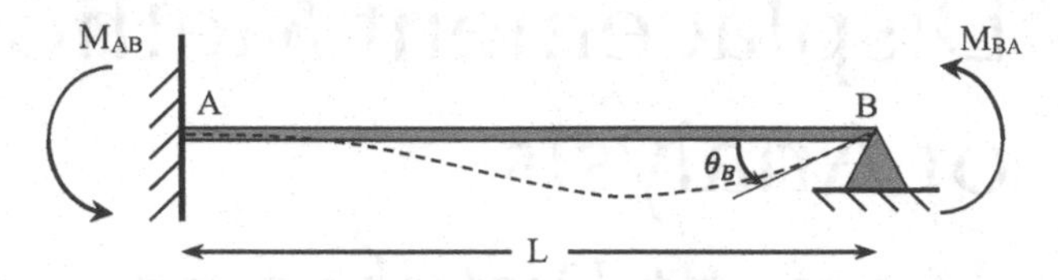

FIGURE 8.1 Beam with moment applied at B.

Figure 8.1 shows a beam with a moment applied at B. It can be proven that

$$M_{BA} = \frac{4EI}{L}\theta_B \tag{8.1}$$

Or,

$$M_{BA} = K\theta_B \tag{8.2}$$

where

$$K = \frac{4EI}{L} \tag{8.3}$$

In equation (8.3), K is the *stiffness factor* for member AB, which is defined as the amount of moment needed at B to induce a unit rotation ($\theta_B = 1$ rad).

In other books (see Chajes (1983) for example) sometimes the stiffness factors are modified based on support conditions, but in this book the authors will advocate using the stiffness factor $K = 4EI / L$ for all members. Using $K = 4EI / L$ for all members will simplify the analysis and provide the same answers.

8.2.2 Carry-Over Factor

In Figure 8.1, it can be proven that the moment induced at A is

$$M_{AB} = \frac{2EI}{L}\theta_B \tag{8.4}$$

From equations (8.1) and (8.4) it can be seen that the carry-over moment, moment induced at A, is $1/2$ of the applied moment at B. This implies that the carry-over factor, which is the ratio of M_A and M_B, is 0.5. Thus, it can be stated that for a beam simply supported at one end and fixed at the other, the CO is 0.5. This concept will be applied in the moment distribution procedure.

8.2.3 Distribution Factor

$$(DF)_{member} = \frac{K_{member}}{\sum K_{member}} \tag{8.5}$$

where $\sum K_{member}$ includes all members connected to the joint considered.

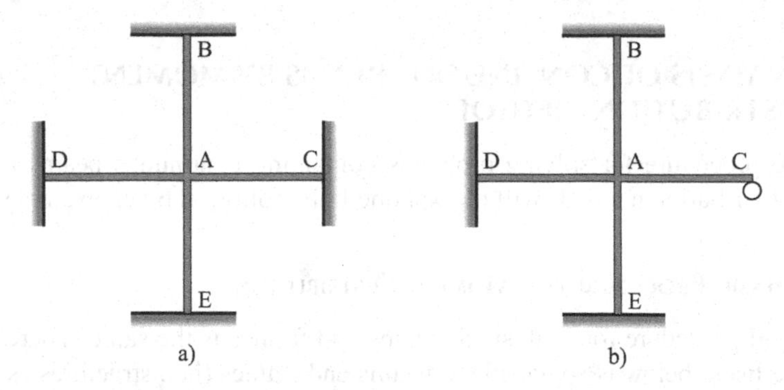

FIGURE 8.2 Concept of distribution factors.

Using Figure 8.2a, the distribution factors can be defined as,

$$(DF)_{AB} = K_{AB} / (K_{AB} + K_{AC} + K_{AD} + K_{AE}) \tag{8.5a}$$

$$(DF)_{AC} = K_{AC} / (K_{AB} + K_{AC} + K_{AD} + K_{AE}) \tag{8.5b}$$

$$(DF)_{AD} = K_{AD} / (K_{AB} + K_{AC} + K_{AD} + K_{AE}) \tag{8.5c}$$

$$(DF)_{AE} = K_{AE} / (K_{AB} + K_{AC} + K_{AD} + K_{AE}) \tag{8.5d}$$

The distribution factor at a fixed support is *zero* because it "absorbs" moments rather than distributing them. Applying equation (8.5) at joint E proves this as can be seen below:

$$DF_{EA} = \frac{K_{EA}}{K_{EA} + \infty} = 0 \qquad \text{(Fixed support)}$$

In theory, a fixed support is "infinitely" stiff because it could take a moment of any size. This makes the denominator of the above equation ∞; therefore, $DF = 0$ for all fixed supports. Similarly, $DF = 1$ for pin and roller support at the end of a beam. Considering joint C in Figure 8.2b, the distribution factor would be calculated as follows:

$$DF_{CA} = \frac{K_{CA}}{K_{CA}} = 1 \qquad \text{(End-pin support)}$$

Here, you can see that since there is only one member attached to end joint C, the stiffness factor is 1. This is true for all pin and roller supports at the ends of continuous beams.

8.3 ANALYSIS OF CONTINUOUS BEAMS BY MOMENT DISTRIBUTION METHOD

The basic procedure for solving problems containing continuous beams using the moment distribution methods will be explained first followed by an example.

8.3.1 Basic Procedure for Moment Distribution

The general procedure for analysis of beams and frames is the same. Therefore, the procedure listed below is applicable to beams and frames (i.e., structures essentially in flexure or bending):

1. Calculate the stiffness factors (K) for each span using the following equation:

$$K = \frac{4EI}{L} \tag{8.3}$$

2. Calculate the distribution factor (DF) for each member using the following relation:

$$DF_{member} = \frac{K_{member}}{\sum K_{member}} \tag{8.5}$$

   where $\sum K_{member}$ includes all members connected to the joint considered.

   Note: Distribution factors that are unknown must be solved using equation (8.5), but those that are known, fixed supports and end-pin supports, can be found immediately.
3. Calculate the fixed-end moments using Table 1A in the Appendix. This step means locking all the joints. The sign convention used is: *clockwise moments and rotations are considered positive*.
4. Set up the moment distribution table by entering the calculated fixed-end moments for each member and the distribution factors for each joint. The table will need to include all the members, and will look similar to the following depending on the number of joints:

Joint	A	B		C		D
Member	AB	BA	BC	CB	CD	DC
DF						
M_F						
Bal						
CO						
Bal						
↓	↓	↓	↓	↓	↓	↓
Final						

5. Start the first cycle of moment distribution by unlocking each joint. Sum up the M_F for all members connected at each joint to get the unbalanced moment ($M_{(unbalanced)}$). Multiply this moment by the respected DF and invert the sign to get the *Bal* for that member. The following relationship can be utilized:

$$Bal = -M_{(unbalanced)} * DF$$

At a pin support, when summing the M_F and *Bal* of each cycle, both members will be delivering the same moment with opposite direction (sign) to the joint. This means the joint is balanced. Fixed supports will have a residual moment.

6. Find the CO by carrying *Bal* values across members (from joint to joint) with a factor of ½. Then get the new unbalancing moments by using the CO. Steps 5 and 6 involve locking and unlocking the joints. The first locking moments are due to fixed moments (caused by the given loads). The successive locking moments are obtained through CO.
7. Continue the balancing until the final unbalance at each joint is about 1% of the initial unbalance moment at any joint.

8.3.2 Example for a Continuous Beam

The procedure discussed above will now be applied to a multi-span continuous beam.

Example 8.3.2.1

Analyze the continuous beam shown in Figure 8.3 using the moment distribution method.

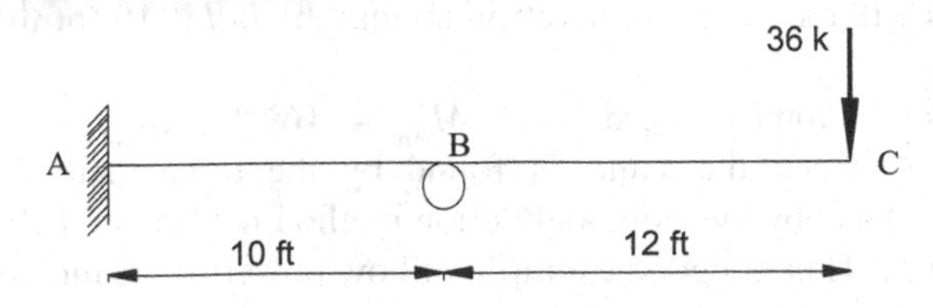

FIGURE 8.3 Continuous beam.

Solution

Step 1

Stiffness factors are not needed here because all the distribution factors are known.

Step 2

$$DF_{AB} = 0 \quad \text{(Fixed support)}$$
$$DF_{BA} = 1 \quad \text{(End-pin support)}$$
$$DF_{BC} = 0 \quad \text{(No Moment is transferred from B to C)}$$

Step 3

$$M_{FAB} = 0, \qquad M_{FBA} = 0, \qquad M_{FBC} = -36*12 = -432^{k-ft}$$

Note: M_{FBC} is negative because the internal moment caused by the loading acts in the counterclockwise direction (opposite to the external moment at that point).

Steps 4–7

The moment distribution table is set up as shown in Table 8.1. First, joints A and B are added to the table along with the corresponding members. Joint C is not part of the table because there is no support. Then, the distribution factors of each member are added based on their support type, and whether they are intermediate or end supports. In this case, joints A and B are end supports because the cantilever portion can be simplified into a moment acting at the BC location. Members AB and BA have no fixed-end moments because there is no loading on the span of the member. The balance for the first cycle in member BA needs to be 432 to satisfy joint equilibrium at joint B. This number can be found either by inspection ($-432 + x = 0$) or by the following standard procedure:

$$Bal = -M_{(unbalanced)} * DF$$

$$Bal_{BA} = -\left(M_{BC} + M_{BA}\right)*1$$

$$Bal_{BA} = -(-432)*1 = 432^{ft-k}$$

The carryover only occurs from joint B to A because moments are not carried to C (cantilever), or from joint A because it is fixed ($DF = 0$). In this problem, the process is repeated only one more time because the *Bal* in the second iteration was all zero. Other problems will require more iteration so that the *Bal* is 1% of the initial unbalanced moment.

The answers found here – $M_{AB} = 216^{ft-k}$ ↻, $M_{BA} = 432^{ft-k}$ ↻, and $M_{BC} = -432^{ft-k}$ ↺ – are the same as found by the force method in Chapter 6 (Example 6.5.2.1) and by the slope-deflection method in Chapter 7 (Example 7.3.1), but with less work. This is a good example of how powerful the moment distribution

TABLE 8.1
Moment distribution table

Joint	A	B	
Member	**AB**	**BA**	**BC**
DF	**0**	**1**	**0**
M_F	0	0	–432
Bal	0	432	0
CO	216	0	0
Bal	0	0	0
Final	**216**	**432**	**–432**

method is; yet the true power of this method will be seen once more complicated examples are solved.

8.4 ANALYSIS OF A CONTINUOUS BEAM WITH SUPPORT SETTLEMENT BY MOMENT DISTRIBUTION METHOD

The procedure for moment distribution discussed in Section 8.3.1 is now applied to a continuous beam with support settlements and no other load. The fixed-end moments at each end are obtained using equation (8.6a) or (8.6b) (Figure 8.4).

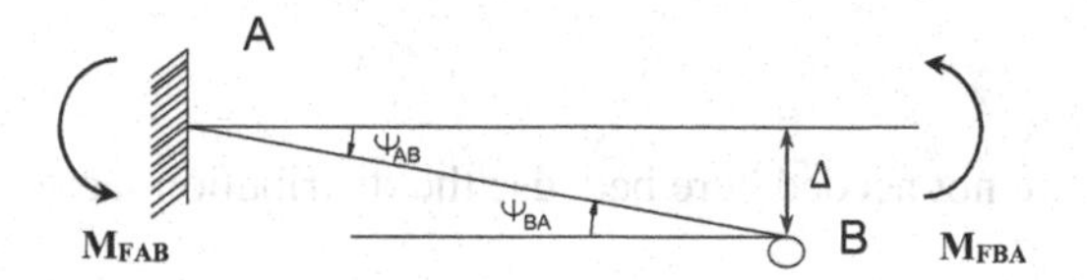

FIGURE 8.4 Effect of displacement at B.

From the right column of Table 1A, when considering the far end pinned, then

$$M_{FAB} = \frac{-3EI\Delta}{L^2} \tag{8.6a}$$

From Table 1A, when considering both ends fixed then,

$$M_{FAB} = M_{FBA} = \frac{-6EI\Delta}{L^2} \tag{8.6b}$$

When solving problems where the far end is pinned, it is possible to take advantage of the right column of Table 1A, which gives fixed-end moments for a structure where the far end is pinned. An example of a far end pinned member is member AB. Although this method can reduce the number of calculations in the moment distribution table, it is not the only way to solve the problem. Alternatively, the standard fixed-end moments can be used for all joints, no matter the given support condition (fixed, pin, or roller). The difference between the two methods is that assuming all joints are fixed (using the left side of Table 1A) can be easier to set up, but it will often involve more iteration in the moment distribution table. In the next example, both methods will be used to show that both methods provide the same answer without a great deal of difference in procedure.

Example 8.4.1

Determine the reactions at the supports of the beam shown in Figure 8.5 by the moment distribution method. Take $E = 29{,}000$ ksi and $I = 446\ in^4$. When the support is displaced by "1".

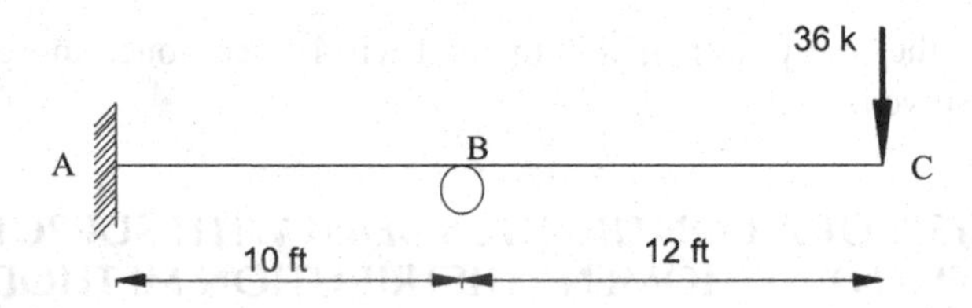

FIGURE 8.5 Statically indeterminate beam with support settlement.

Solution

Method 1

Step 1

Stiffness factors are not needed here because the distribution factors are known.

Step 2

$$DF_{AB} = 0 \quad \text{(Fixed support)}$$

$$DF_{BA} = 1 \quad \text{(End-pin support)}$$

$$DF_{BC} = 0 \quad \text{(No Moment is transferred from B to C)}$$

Step 3

Find M_{FAB} due to settlement downward of support B

$$\rightarrow M_{FAB} = -\frac{3EI\Delta}{L^2} = -\frac{3(4176000)(0.0215)\left(\frac{1}{12}\right)}{(10)^2} = -224.46^{k-ft}$$

$$M_{FBC} = -36(12) = -432^{k-ft}$$

Note: M_{FBC} is negative because the internal moment caused by the loading acts in the counterclockwise direction (opposite to the external moment at that point).

Steps 4 – 7

Now the moment distribution table is filled in using the distribution factors and fixed-end moments found in steps 1 through 3. In this moment distribution table, the unbalanced moments are equal to the fixed-end moments at each joint. Just as was done in the first problem, the unbalanced moment is multiplied by the respected *DF* and the *Bal* is found. Only one carryover takes place, which is from BA to AB. After only two cycles the moment distribution table is finished because all *Bal* values are zero (Table 8.2).

TABLE 8.2
Moment distribution table (method 1)

Joint	A	B	
Member	AB	BA	BC
DF	0	1	0
M_F	–224.46	0	–432
Bal	0	432	0
CO	216	0	0
Bal	0	0	0
ΣM	**–8.46**	**432**	**–432**

The answers found here $-M_{AB} = -8.46^{ft-k}$ ↺, $M_{BA} = 432^{ft-k}$ ↻, and $M_{BC} = -432^{ft-k}$ ↺ – are the same as found by the force method in Chapter 6 (Example 6.5.2.1) and by the slope-deflection method in Chapter 7 (Example 7.3.1), but with less work.

Method 2

Step 1

Stiffness factors are not needed because the distribution factors are known.

Step 2

$$DF_{AB} = 0 \quad \text{(Fixed support)}$$

$$DF_{BA} = 1 \quad \text{(End-pin support)}$$

$$DF_{BC} = 0 \quad \text{(No Moment is transferred from B to C)}$$

Step 3

$$M_{FAB} = M_{FBA} = -\frac{6EI\Delta}{L^2} = -\frac{6(4176000)(0.0215)\left(\frac{1}{12}\right)}{(10)^2} = -488.92^{k-ft}$$

Step 4

Since the left column fixed-end moments are used, member BA now has the same moment as member AB. The unbalanced moment for joint B is found by summing up M_{FBA} and M_{FBC}. This value is then multiplied by the distribution factor of members BA and BC as seen below:

$$Bal = -M_{(unbalanced)} * DF$$

$$Bal_{BA} = -\left(M_{BC} + M_{BA}\right) * 1$$

$$Bal_{BA} = -(-432 + (-448.92)) * 1 = 880.92^{ft-k}$$

Next, the carry-over factor ($\frac{1}{2}$) is applied from joint B to Joint A, as indicated by the arrows. From here all balances are zero and so the moments are summed up and the table is complete (Table 8.3).

When comparing the two methods, the final moments are exactly the same, and found in only two cycles. This proves that both methods can be used for a given problem based on preference. The authors of this book prefer method 2 due to the simplicity of finding the fixed-end moments. It should also be noted that the results are **identical** to those obtained using the slope-deflection method in Example 7.4.1 which can be seen in the chapter summary in Section 7.7.

TABLE 8.3
Moment distribution table (method 2)

Joint	A	B	
Member	AB	BA	BC
DF	0	1	0
M_F	–448.92	–448.92	–432
Bal	0	880.92	0
CO	440.46	0	0
Bal	0	0	0
ΣM	–8.46	432	–432

8.5 APPLICATION OF MOMENT DISTRIBUTION TO ANALYSIS OF FRAMES WITHOUT SIDESWAY

The analysis of frames without sidesway is similar to that of continuous beams. The procedure described in Section 8.3 will be used to analyze frames without sidesway. An example is discussed below to illustrate this concept.

EXAMPLE 8.5.1

Determine the moments at each joint of the frame shown in Figure 8.6a by the moment distribution method. $E = 29{,}000$ ksi, $A = 16$ in^2, and $I = 446$ in^4 for all members.

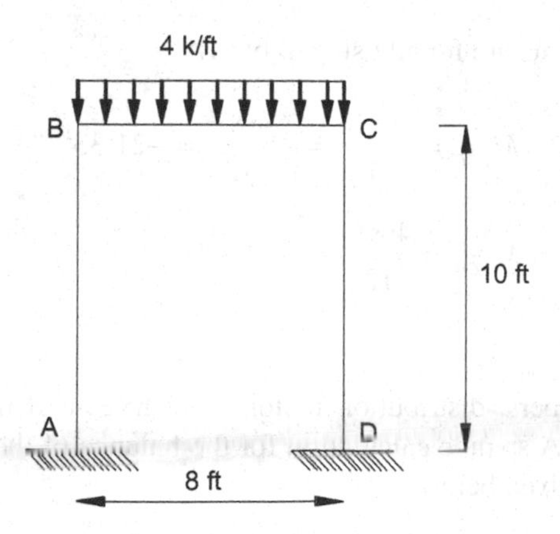

FIGURE 8.6a Indeterminate frame (no sidesway).

Solution

Step 1

$$K_{AB} = \frac{4EI}{10},\ K_{BC} = \frac{4EI}{8},\ \text{and}\ K_{CD} = \frac{4EI}{10}$$

Step 2

In this problem we must find the distribution factors for the members at joints B and C using equation (8.5) because they are unknown. At joint B,

$$\rightarrow DF_{BA} = \frac{\frac{4EI}{10}}{\frac{4EI}{10} + \frac{4EI}{8}} = 0.444$$

$$\rightarrow DF_{BC} = 1 - 0.444 = 0.556$$

Similarly, at joint C,

$$\rightarrow DF_{CB} = \frac{\frac{4EI}{8}}{\frac{4EI}{10} + \frac{4EI}{8}} = 0.556$$

$$\rightarrow DF_{CD} = 1 - 0.556 = 0.444$$

Step 3

The fixed-end moments are found using Table 1A.

$$M_{FBC} = -\frac{wL^2}{12} = -\frac{4(8)^2}{12} = -21.33^{k-ft}$$

$$M_{FCB} = \frac{4(8)^2}{12} = 21.33^{k-ft}$$

Steps 4–7

All the joints, members, distribution factors, and fixed-end moments are filled in based on steps 1–3. A sample calculation for the balance of the first cycle for members BA and BC is given below:

$$Bal = -M_{(unbalanced)} * DF$$

$$Bal_{BA} = -(M_{BC} + M_{BA}) * 0.444$$

$$Bal_{BA} = -(-21.33 + 0) * 0.444 = 9.48^{ft-k}$$

$$Bal_{BC} = -(M_{BC} + M_{BA}) * 0.556$$

$$Bal_{BC} = -(-21.33 + 0) * 0.556 = 11.85^{ft-k}$$

Carry-over is applied between joints B and C, from B to A, and from C to D. This process is repeated until the balance is about 1% of the original unbalanced moment. Lastly, values for the internal end moments of each member are found by summing up all the entries in each member's column starting with the fixed-end moment (Table 8.4).

The answers for the reactions at the base would be the moments, $M_{AB} = 6.55^{k-ft}$ ↻ and $M_{DC} = -6.55^{k-ft}$ ↺ (Figure 8.6b). These values of end moments are very similar to the results of slope-deflection method.

Note: M_{AB} is the internal moment at joint A in member AB and is also the reaction at the support, M_A. M_A has the same magnitude and direction as M_{AB}.

The vertical reactions at the base can be found using basic statics because the loading is symmetrical (half of the total distributed load force applied to each support acting upward). Also, the horizontal forces at the base can be found in the following fashion:

TABLE 8.4
Moment distribution table

Joint	A	B		C		D
Member	AB	BA	BC	CB	CD	DC
DF	0	0.444	0.556	0.556	0.444	0
M_F	0	0	−21.33	21.33	0	0
Bal	0	9.48	11.85	−11.85	−9.48	0
CO	4.74	0	−5.925	5.925	0	−4.74
Bal	0	2.633	3.292	−3.292	−2.633	0
CO	1.317	0	−1.646	1.646	0	−1.317
Bal	0	0.731	0.914	−0.914	−0.731	0
CO	0.366	0	−0.457	0.457	0	−0.366
Bal	0	0. 203	0.254	−0.254	−0.203	0
CO	0.102	0	−0.127	0.127	0	−0.102

Bal	0	0.056	0.071	–0. 071	–0.056	0
CO	0.028	0	–0.036	0.036	0	–0.028
Bal	0	0.016	0.02	–0.02	–0.016	0
Σ*M*	**6.55**	**13.12**	**–13.12**	**13.12**	**–13.12**	**–6.55**

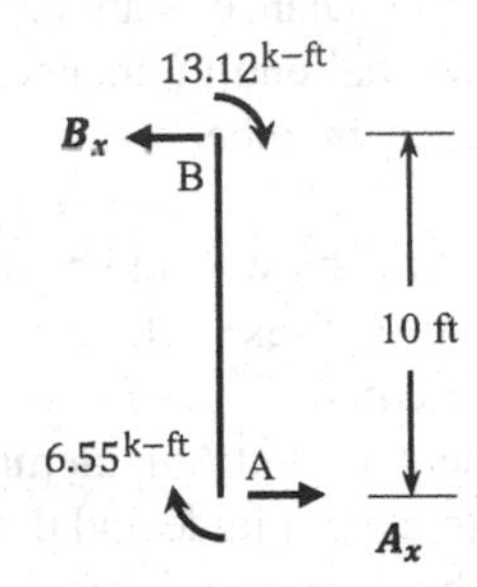

FIGURE 8.6b Member AB.

8.6 APPLICATION OF MOMENT DISTRIBUTION TO ANALYSIS OF FRAMES WITH SIDESWAY

In this section, the basic concept involved in the analysis of a frame with sidesway by moment distribution will be discussed followed by an example.

8.6.1 Basic Concepts: Application of Moment Distribution to Analysis of Frames with Sidesway

To solve a frame with sidesway, the principle of superposition will be utilized (Figure 8.7). This analysis involves two steps: (1) analyze the frame with sidesway being restrained and with the applied loading (see Figures 8.7b, and 2), analyze the frame with only sidesway and no applied loads (Figure 8.7c). In both steps the moment distribution table will need to be used.

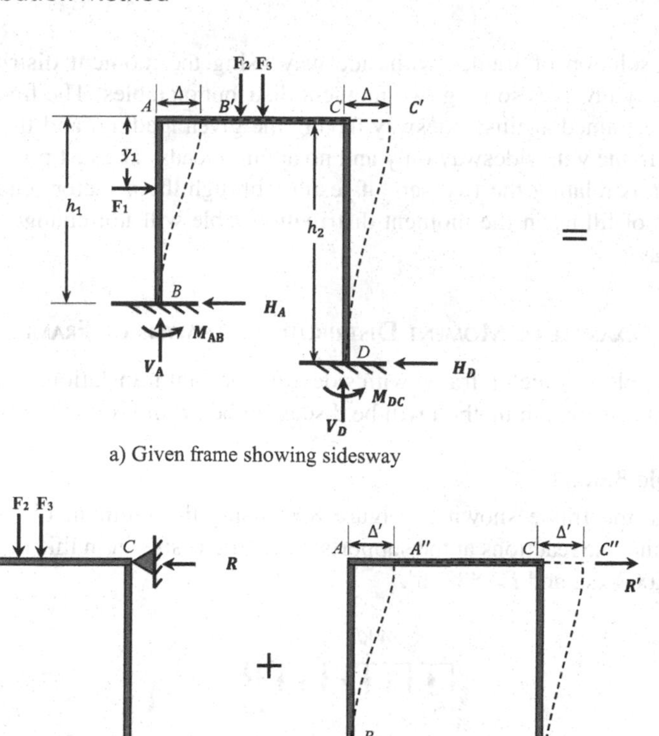

a) Given frame showing sidesway

b) Sidesway restrained

c) Sidesway allowed

FIGURE 8.7 Principle of superposition applied to a frame with sidesway.

The basic superposition equation for the moment can be derived as

$$M_{NF} = M'_{NF} + kM''_{NF} \tag{8.7}$$

where

$$k = R / R' \tag{8.8}$$

N = Near joint of a member
M = Far joint of a member

The single prime denotes the moments on the restrained frame (Figure 8.7b), and the double prime denotes the moments due to sidesway only (Figure 8.7c). The expressions for sidesway moments are given in equations (8.6a) and (8.6b).

The solution of frames with sidesway using the moment distribution methods essentially involves solving two moment distribution tables. The first table is for the frame restrained against sidesway having the given loading, and the second table is for the frame with sidesway only and no applied loads. The last portion of the superposition is relating the two sets of results through the k factor (equation 8.8). The process of filling in the moment distribution table will not change for frames with sidesway.

8.6.2 Example of Moment Distribution: Analysis of Frames with Sidesway

An example in which a frame with sidesway, or joint translation, is solved using the moment distribution method will be discussed below in Example 8.6.2.1.

Example 8.6.2.1

Analyze the frame shown in Figure 8.8a using the moment distribution method. Determine the reactions at the supports of the frame shown in this figure. $A = 100$ in^2, $E = 29{,}000$ ksi, and $I = 833$ in^4.

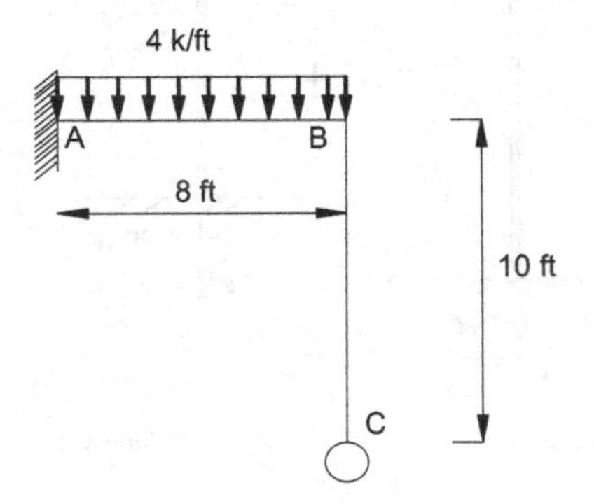

FIGURE 8.8a Frame with sidesway.

Step 1

Here we need to calculate the stiffness factors of the members connected to joint B, because we do not know DF_{BA} or DF_{BC}.

$$K_{AB} = K_{BA} = \frac{4EI}{8} = 0.5EI$$

$$K_{BC} = K_{CB} = \frac{4EI}{10} = 0.4EI$$

Step 2

The distribution factors for members BA and BC are calculated below. DF_{BC} is found by $1 - DF_{BA}$ because the sum of the distribution factors at joint B must be 1. Also, since C is an end-pin support, DF_{CB} is 1.

$$DF_{BA} = \frac{0.5EI}{0.5EI + 0.4EI} = 0.556$$

$$DF_{BC} = 1 - 0.556 = 0.444$$

$$DF_{CB} = 1 \quad \text{(end-pin support)}$$

Step 3 – Restrained frame

The restrained frame shown in Figure 8.8b includes an artificial support to inhibit sidesway at joint C. *R* is the reaction force at the artificial joint, C, due to the given loading. This is the first part of the principle of superposition applied to the frame.

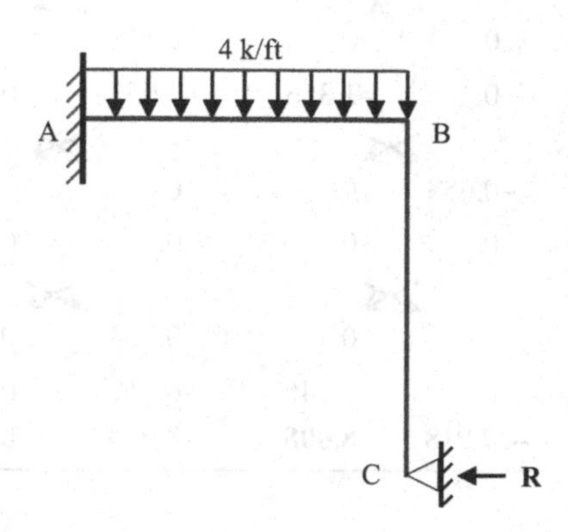

FIGURE 8.8b Restrained frame (no sidesway).

The fixed-end moments for member AB are calculated using Table 1A and are shown below. There are no fixed-end moments on member BC because there is no external loading on the member's span.

$$M_{FAB} = -\frac{4*8^2}{12} = -21.33^{k-ft} \quad and \quad M_{FBA} = \frac{4*8^2}{12} = 21.33^{k-ft}$$

Step 4

Table 8.5, the moment distribution table for the restrained frame, is set up using the distribution factors and the fixed-end moments found in the first three steps. Then, iteration of *Bal* and *CO* are carried out, ensuring that joint equilibrium is satisfied in each cycle. The carry-over factor of $\frac{1}{2}$ is applied between joints where possible. Summing up each column yields the moments at each end of the members.

The results obtained here are, $M'_{AB} = -27.918^{k-ft}$ ↺,

$$M'_{BA} = 8.008^{k-ft} ↻, \text{ and } M'_{BC} = -8.008^{k-ft} ↺.$$

TABLE 8.5
Moment distribution table for the restrained frame

Joint	A	B		C
Member	AB	BA	BC	CB
DF	**0**	**0.556**	**0.444**	**1**
M_F	−21.33	21.33	0	0
Bal	0	−11.859	−9.471	0
CO	−5.930	0	0	−4.736
Bal	0	0	0	4.736
CO	0	0	2.368	0
Bal	0	−1.316	−1.051	0
CO	−0.658	0	0	−0.526
Bal	0	0	0	0.526
CO	0	0	0.263	0
Bal	0	−0.146	−0.117	0
Final	**−27.918**	**8.008**	**−8.008**	**0**

Calculate R:

Using the internal moments found for the restrained frame in Table 8.5, we can now find the force R.

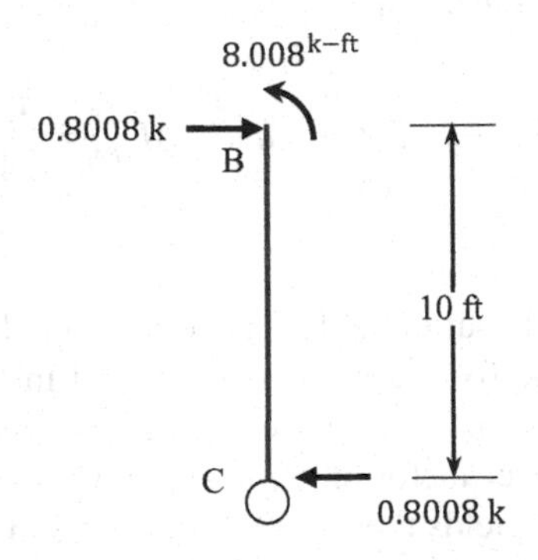

FIGURE 8.9 Calculation of R using member BC.

Using Figure 8.9, the following calculation can be made:

$$Required\,couple = \frac{8.008^{ft-k}}{10\,ft} = 0.8008\,k$$

In order to satisfy equilibrium on this member, the couple must be acting in the opposite direction as the M_{BC}, therefore:

$$R = 0.8008\,k$$

Note: R is positive because it is acting in the direction assumed (Figure 8.8).

8.6.3 Frame with Sidesway and Artificial Joint Removed

Again, the fixed-end moments on the members need to be found. Member BC will have a fixed-end moment due to deflection at C. Typically this would involve using Table 1A found in the appendix, but if the columns have the same displacement (Δ'), *E*, *I*, and *L*, then an arbitrary value can be used with the correct direction. In this problem, there is only one column and one moment so no problems will be encountered using an arbitrary moment of 100^{k-ft} applied clockwise at B (Figure 8.10).

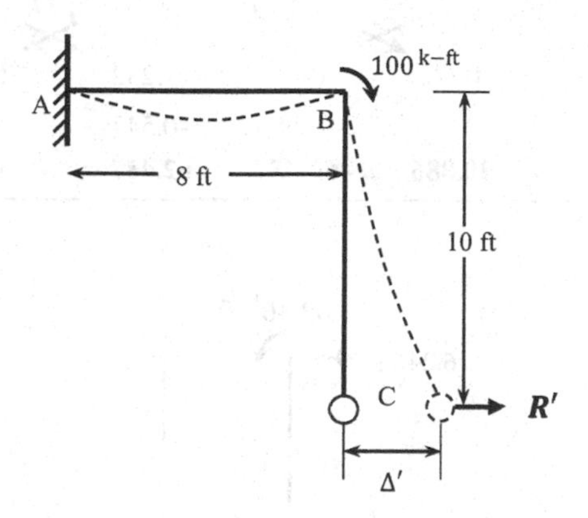

FIGURE 8.10 Frame with sidesway and artificial joint removed.

To set up Table 8.6, all we need are the distribution factors and the single M_F for member BC. Then, the typical procedure of *Bal* and *CO* are used to fill in the rest of the table and find the end moments at each member.

The results obtained here are, $M''_{AB} = -30.886^{k-ft}$ ↺,

$$M''_{BA} = -62.457^{k-ft} \circlearrowleft, \text{ and } M''_{BC} = 62.457^{k-ft} \circlearrowright$$

Calculate R':

Now we will find the value of R', which is the amount of force required to make the displacement, Δ', at joint C using the moments found in Table 8.6. This is the second part of the principle of superposition for the frame (Figure 8.10).

TABLE 8.6
Moment distribution table for the frame with sidesway

Joint	A	B		C
Member	AB	BA	BC	CB
DF	0	0.556	0.444	1
M_F	0	0	100	0
Bal	0	–55.6	–44.4	0
CO	–27.8	0	0	–22.2
Bal	0	0	0	22.2
CO	0	0	11.1	0
Bal	0	–6.172	–4.928	0
CO	–3.086	0	0	–2.464
Bal	0	0	0	2.464
CO	0	0	1.232	0
Bal	0	–0.685	–0.547	0
Final	**–30.886**	**–62.457**	**62.457**	**0**

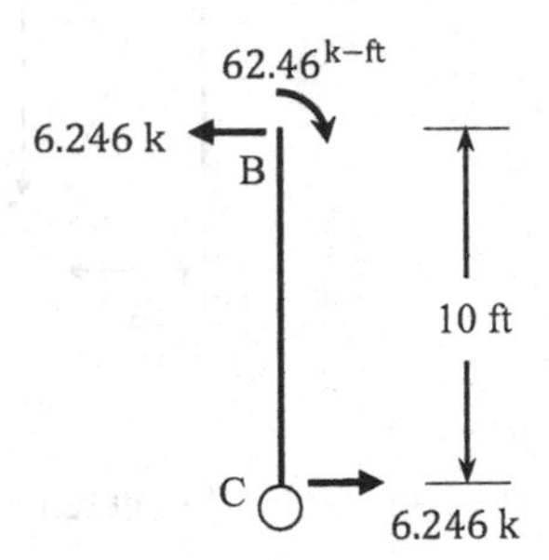

FIGURE 8.11 Calculation of $\boldsymbol{R'}$ using member BC.

Using Figure 8.11, the following calculation can be made:

$$Required\ couple = \frac{62.46^{ft-k}}{10ft} = 6.246k$$

In order to satisfy equilibrium on this member, the couple must be acting in the opposite direction as the M_{BC}, therefore:

$$R' = 6.246\,k$$

Note: R' is positive because it acts in the direction of R' that is assumed in Figure 8.10 are computed as:

Using the ratio of R/R' we can complete the superposition and find the amount of moment that needs to be added to or subtracted from the original moments found for the restrained frame.

$$M_{AB} = M'_{AB(restrained)} + \frac{R}{R'} M''_{AB(sidesway)}$$

$$M_{AB} = -27.918 + \frac{0.8008}{6.246}(-30.886) = -32^{k\text{-}ft}$$

$$M_{BA} = 8.008 + \frac{0.8008}{6.246}(-62.46) = 0^{k\text{-}ft}$$

$$M_{BC} = -8.008 + \frac{0.8008}{6.246}(62.46) = 0^{k\text{-}ft}$$

The reactions at A (A_x and A_y) and C (C_y) are calculated from principles of statics (Figure 8.12):

$$\overset{+}{\rightarrow} \Sigma F_x = 0 \rightarrow A_x = 0$$

$$\overset{+}{\circlearrowright} \Sigma M_A = 0 : -32 + 32 * 4 - C_y * 8 = 0$$

$$C_y = 12\boldsymbol{k}(\uparrow)$$

$$+\uparrow \Sigma F_y = 0 : A_y + 12 - 32 = 0$$

$$A_y = 20\,\boldsymbol{k}(\uparrow)$$

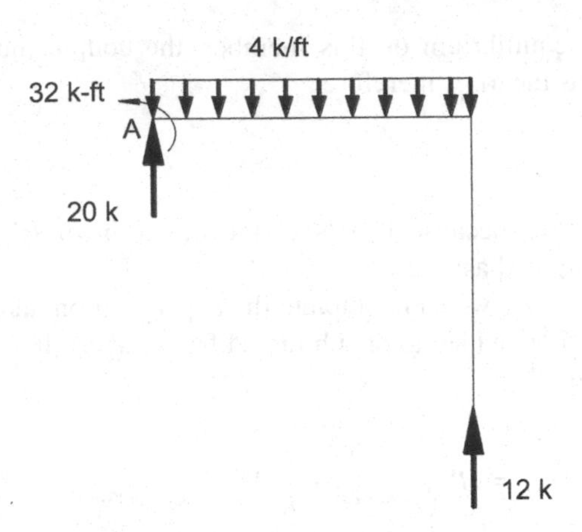

FIGURE 8.12 Final reactions for the frame.

8.7 SUMMARY

A concise description of the concept of the moment distribution is presented in this chapter. This is followed by few problems dealing with continuous beams, frame without joint movements (sidesway), and with joint movements. It should be noted that moment distribution method directly gives moments.

Table 8.7 shows a summary of the solutions to example problems from Chapters 6 to 8. The answers for each method are very close if not exactly the same, showing that any method may be used to analyze an indeterminate structure. The moment

TABLE 8.7
Comparison of example problem solutions from Chapters 1 to 3

	Chapter 6 (Force Method)	Chapter 7 (Slope-Deflection)	Chapter 8 (Moment Distribution)
Beam	$B_y = 100.8k \uparrow$ $A_y = -64.8k \downarrow$ $M_A = 216^{k\text{-}ft}$ ↻	$B_y = 100.8k \uparrow$ $A_y = -64.8k \downarrow$ $M_{AB} = 216^{k\text{-}ft}$ ↻	$B_y = 100.8k \uparrow$ $A_y = -64.8k \downarrow$ $M_{AB} = 216^{k\text{-}ft}$ ↻
Beam with settlement		$B_y = 78.356k \uparrow$ $A_y = -43.356k \downarrow$ $M_{AB} = 8.44^{k\text{-}ft}$ ↺	$B_y = 78.354k \uparrow$ $A_y = -43.354k \downarrow$ $M_{AB} = 8.46^{k\text{-}ft}$ ↺
Frame no sidesway		$M_{AB} = 6.573^{k\text{-}ft}$ ↻ $M_{AB} = 6.573^{k\text{-}ft}$ ↺ $A_y = D_y = 16k \uparrow$	$M_{AB} = 6.55^{k\text{-}ft}$ ↻ $M_{AB} = 6.55^{k\text{-}ft}$ ↺ $A_y = D_y = 16k \uparrow$
Frame with sidesway	$A_y = 20k \uparrow$ $C_y = 12k \downarrow$ $M_A = 32^{k\text{-}ft}$ ↺	$A_y = 20k \uparrow$ $C_y = 12k \downarrow$ $M_{AB} = 32^{k\text{-}ft}$ ↺	$A_y = 20k \uparrow$ $C_y = 12k \downarrow$ $M_{AB} = 32^{k\text{-}ft}$ ↺

distribution method was the only method that showed slight differences. These can usually be minimized with more iteration.

PROBLEMS FOR CHAPTER 8

Analyze Problems 8.1 to 8.3 using the moment distribution method.

Problem 8.1 Solve *Problem 7.1* by using the moment distribution method.

(*Problem 7.1 repeated:* Determine the reactions at the supports of the beam shown in this figure. EI is constant)

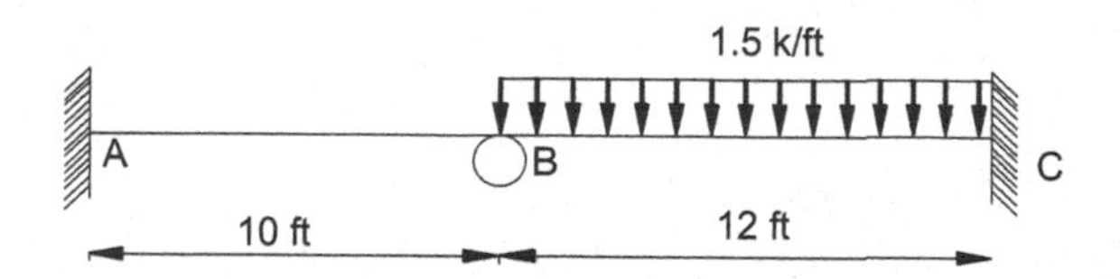

Problem 8.2 Solve *Problem 7.2* by using the moment distribution method

(*Problem 7.2 repeated:* Determine the reactions at the supports of the frame shown in this figure. $A = 100$ in^2, $E = 29{,}000$ ksi, and $I = 833$ in^4).

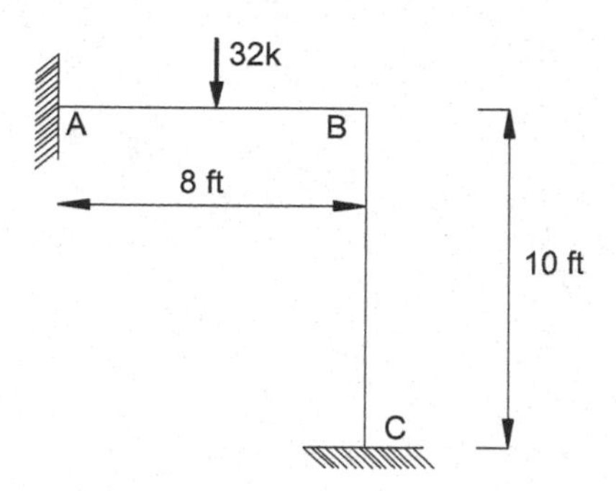

Problem 8.3 Solve *Problem 7.3* by using the moment distribution method

(*Problem 7.3 repeated:* Determine the reactions at the supports of the frame shown in this figure. $A = 100$ in^2, $E = 29{,}000$ ksi, and $I = 833$ in^4).

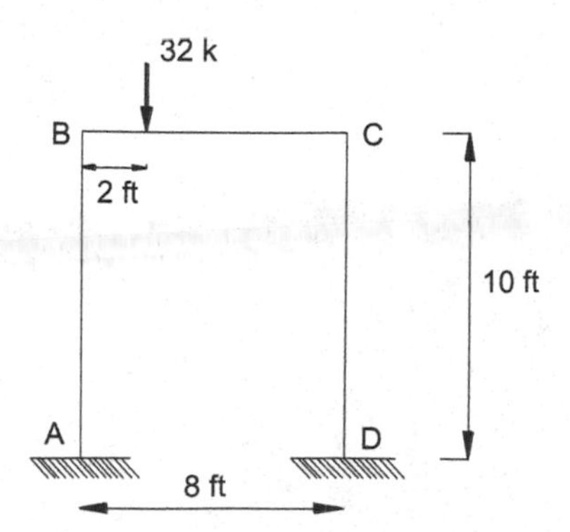

9 Direct Stiffness Method
Application to Beams

9.1 BASIC CONCEPTS OF THE STIFFNESS METHOD

Stiffness method is the most commonly used method used for analysis of structures. Almost all the computer codes written to analyze structures use the stiffness method. One of the reasons for its wide use is that the general procedure of the stiffness method can be applied to any type of structure, for example, beam, frame, truss, or any structure for that matter. Also, it is very easy to use and can be coded for analysis of entire structures.

9.2 KINEMATIC INDETERMINACY

A structure's kinematic indeterminacy (KI) must be established before solving a problem by the stiffness method. Again, kinematic indeterminacy is defined as the total number of degrees of freedom for all the joints in a given structure. Another method similar to the stiffness methods is the flexibility method. The flexibility method is a matrix equivalent of the force method.

9.3 RELATION BETWEEN STIFFNESS METHOD AND DIRECT STIFFNESS METHOD

While the stiffness method and the direct stiffness method are essentially the same, well-known researchers have drawn a distinction between them (Weaver and Gere, 1990). Although the distinction is slight, it is important that this is explained, especially for the undergraduate students in civil engineering for whom this book is aimed at. In the stiffness method, the elements of a *stiffness matrix are derived* from the basic principles of engineering mechanics corresponding to the unknown displacements in the structure. In the case of the direct stiffness matrix, the *standard stiffness matrix* for each element (whether beam element, truss element, or frame element) is used to assemble a structure's stiffness matrix. This matrix is then used to solve for displacements. Thus, the direct stiffness method is more mechanical, to put it in plain terms, and is very easy to use. For this reason, the direct stiffness method is very popular and widely used for the analysis of any type of structure. The details of the direct stiffness method will be discussed in the text and also applied to beam, frame, and truss structures with specific examples of each in this chapter, Chapter 10, and Chapter 11, respectively.

DOI: 10.1201/9781003246633-11

9.4 DERIVATION/EXPLANATION OF THE BEAM-ELEMENT STIFFNESS MATRIX

A typical beam element, or member, is shown in Figure 9.1. As can be seen in Figure 9.1, the action $\{A\}$ and the displacement $\{D\}$ are shown at the ends of a typical beam element. $\{D\}$ represents a generalized displacement (translation or rotation) and $\{A\}$ represents a generalized force (force or moment). In the stiffness method, all loads will be distributed to the nodes. An assembly of the various actions of a structure will constitute an action vector. Similarly, an assembly of the various displacements of a structure will constitute a displacement vector. The reactions at the supports will also be considered part of the generalized action vector. In Figure 9.1, the actions and displacements share a common number. To simplify things, one number and symbol are often written to represent both entities (see Figure 9.3).

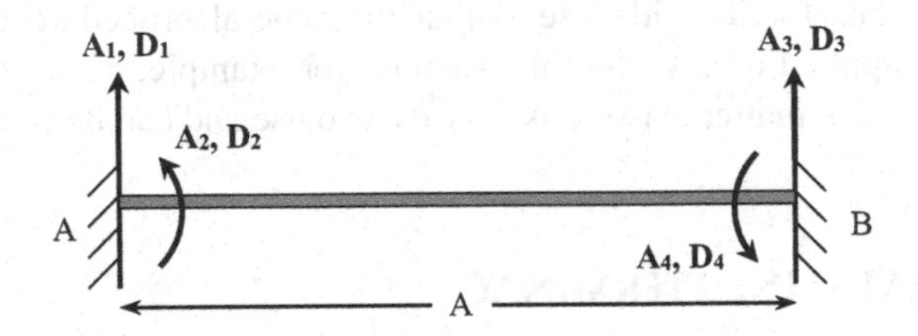

FIGURE 9.1 Typical beam element.
Note: The sign convention used in the stiffness method follows the right hand rule. This means curling the fingers on your right hand so that they point in the direction of the moment or rotation, if your thumb points up or at you, then it is positive. Likewise, if it points down or away from you, then it is negative.

The basic definition of stiffness can easily be obtained from the following equation which is the basic relationship used in the stiffness method:

$$\{A\} = [K]\{D\} \tag{9.1}$$

where $\{A\}$ = Action vector
$\{D\}$ = Displacement vector
$[K]$ = Stiffness matrix

From equation (9.1), if $\{D\} = 1$ then $[K] = \{A\}$. This implies that the force required to cause a unit displacement is the stiffness. This is a very important definition. This basic relation is used in the analysis of every structure. If there is only one displacement, then $[K]$ will be an element instead of a matrix. $[K]$ will be different from element to element when the length (L), material (E), or cross-section (I) are changed .

The matrix $[K]$, used to relate the actions at joint A (near end) and joint B (far end) to the displacements at joint A and B for a beam element can be expressed as (Weaver and Gere, 1990) follows:

$$[K]=\begin{array}{c}\begin{array}{cccc} D_1 & D_2 & D_3 & D_4 \end{array}\\ \left[\begin{array}{cccc} \dfrac{12EI}{L^3} & \dfrac{6EI}{L^2} & -\dfrac{12EI}{L^3} & \dfrac{6EI}{L^2} \\ \dfrac{6EI}{L^2} & \dfrac{4EI}{L} & -\dfrac{6EI}{L^2} & \dfrac{2EI}{L} \\ -\dfrac{12EI}{L^3} & -\dfrac{6EI}{L^2} & \dfrac{12EI}{L^3} & -\dfrac{6EI}{L^2} \\ \dfrac{6EI}{L^2} & \dfrac{2EI}{L} & -\dfrac{6EI}{L^2} & \dfrac{4EI}{L} \end{array}\right]\end{array}\begin{array}{c} \\ A_1 \\ A_2 \\ A_3 \\ A_4 \end{array} \tag{9.2}$$

Note: $[K]$ is a symmetric matrix.

The action and displacement vectors for the beam element shown in Figure 9.1 can be expressed as,

$$\{A\}=\begin{Bmatrix} A_1 \\ A_2 \\ A_3 \\ A_4 \end{Bmatrix} \quad and \quad \{D\}=\begin{Bmatrix} D_1 \\ D_2 \\ D_3 \\ D_4 \end{Bmatrix} \qquad \text{(9.3a and 9.3b)}$$

The $[K]$ matrix in equation (9.2) is written in structure coordinates (global x, y, z axes), which is the same as member axes for a beam element. The member axis changes with respect to the angle of the member. For example, the member axis of a column (vertical member) would be rotated 90° counterclockwise from the structure axis. Frame and truss elements that are not horizontal will have different member and global axis. Hence, $[K]$ can also be called $[K_{MSi}]$, which means stiffness matrix of the element in structure (global) coordinates. It is to be noted that the way the stiffness matrix $[K]$ is written in equation (9.2) has to properly correspond with the action and displacement vectors in equation (9.3). As can be seen by the markings D_1 through D_4 in equation (9.2), the rows and columns of $[K]$ correspond to the translational and rotational displacements of a given member. The pattern used in equation (9.2) is not arbitrary. The first row corresponds to the translational displacement at the near (typically left) end of the member and the second row corresponds to the rotational displacement at the near end of the member. The third and fourth rows correspond to the translation and rotation of the far (typically right) end of the member, respectively. This pattern must be followed for all members. Otherwise, $[K]$ will be wrong and erroneous results will be obtained. The displacement vector has to consist of translations and rotations at the near end followed by translations and rotations at the far end. The same order is to be followed for the corresponding actions when determining the action vector.

Another point to be noted is that some of the displacements in the displacement vector will be unknown because they are free to displace. These will be called $\{D_F\}$ because they are the free displacements. On the other hand, some displacements will be known because they are zero, or move a certain amount (support settlement). These are designated as $\{D_R\}$ because they are the restrained displacements. The free displacements are the ones to be obtained. The actions corresponding to the free displacements are denoted by $\{A_F\}$. They are known because they are the given applied loads. The unknown required forces (such as reactions) are designated as $\{A_R\}$ because they correspond to the restrained displacements.

To summarize, we want to find the unknown free displacements $\{D_F\}$ and the unknown required forces $\{A_R\}$ using the known restrained displacements $\{D_R\}$ and the known given loads $\{A_F\}$.

9.4.1 Global/Structure Stiffness Matrix

In addition to the element stiffness matrix $[K]$, there is the structure stiffness matrix $[K_J]$, which links each individual member to the whole structure. To make the member stiffness matrix for a given structure, the following formula will need to be used:

$$K_{Jij} = \sum_{1}^{n} K_{A_i - D_j} \tag{9.4}$$

where n = Number of terms with the same i-j location
$K_{A_i - D_j}$ = Value in the element stiffness matrix $[K]$ corresponding to A_i and D_j.

This formula is used to find each term of $[K_J]$ using the i-j location in the element stiffness matrix.

9.5 Application of the Direct Stiffness Method to a Continuous Beam

Here, the basic analysis procedure of the direct stiffness method for continuous beams will be explained followed by an example.

9.5.1 Basic Procedure of the Direct Stiffness Method for Beams

1. Number the joints.
2. Number and define the members (with respect to how the member is connected and to which joints). Follow the same order when defining the displacements and the corresponding actions.
3. Determine kinematic indeterminacy (KI) and identify the unknown displacements for the given structure. Then, number the displacements starting from the free (unknown) displacements followed by the restrained displacements. Within the free displacements any order can be followed.
4. Write down the element stiffness matrix $[K]$ from equation (9.2) connecting the action vector and the displacement vector (from equation 9.3) for all the members in the given structure. Again, make sure that the translational

displacements are to be followed by rotational displacements (follow the same corresponding order for action vector as well).

5. Assemble the structure stiffness matrix $[K_J]$ by combining the elements of the same kind using the following equation:

$$K_{Jij} = \sum_{1}^{n} K_{A_i - D_j} \tag{9.4}$$

where n = Number of terms with the same *i-j* location
$K_{A_i - D_j}$ = Value in the element stiffness matrix $[K]$ corresponding to A_i and D_j

$K_{A_i - D_j}$ corresponds to all elements of the member stiffness matrix $[K]$ that relate to a specific action, A_i, and displacement, D_j. The above equation generates all the elements of the structure stiffness matrix $[K_J]$ by combining all the elements corresponding to suffix *i-j* for all the members.

If the numbering of the displacements is done starting from the free displacements, the constitution of $[K_J]$, the joint structure stiffness matrix, as it relates to structure action and displacement will be as shown in equation (9.5).

$$\begin{Bmatrix} A_F \\ A_R \end{Bmatrix} = \begin{bmatrix} K_{FF} & K_{FR} \\ K_{RF} & K_{RR} \end{bmatrix} \begin{Bmatrix} D_F \\ D_R \end{Bmatrix} \tag{9.5}$$

where K_{FF} = Portion of the structure stiffness matrix containing the known actions $\{A_F\}$ and the corresponding free displacements $\{D_F\}$
K_{FR} = Portion of the structure stiffness matrix containing the known actions $\{A_F\}$ and the restrained displacements $\{D_R\}$
K_{RF} = Portion of the structure stiffness matrix containing the unknown forces $\{A_R\}$ and the free displacements $\{D_F\}$
K_{RR} = Portion of the structure stiffness matrix containing the unknown forces $\{A_R\}$ and the restrained displacements $\{D_R\}$

6. The unknown displacements $\{D_F\}$ can be obtained from the following equation (which is derived from equation 9.5).

$$\{D_F\} = [K_{FF}]^{-1} \{A_{FN}\} \tag{9.6}$$

where $\{A_{FN}\}$ denotes the net actual and equivalent joint loads corresponding to free displacements. Equivalent joint loads are member loads that are distributed to the joints.

7. The unknown reactions $\{A_R\}$ can be obtained from the following equation:

$$\{A_R\} = [K_{RF}]\{D_F\} - \{A_{RN}\} \tag{9.7}$$

where $\{A_{RN}\}$ denotes the net actual and equivalent joint loads corresponding to restrained displacements. Using the above equation, one can easily obtain $\{A_R\}$ knowing $[K]$ and $\{D\}$ obtained in previous steps.

9.5.2 Example of a Continuous Beam Using the Stiffness Method

A continuous beam is solved below using the stiffness method described above.

Example 9.5.2.1

Determine the reactions at the supports of the beam shown in Figure 9.2 using the stiffness method. $E = 29{,}000$ ksi and $I = 446$ in^4.

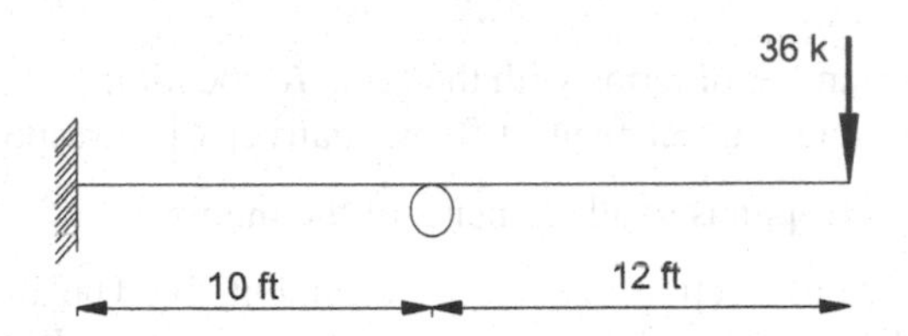

FIGURE 9.2 Continuous beam.

Solution

Steps 1–3

Figure 9.3 shows the beam after executing steps 1 through 3. Member 1 is defined as node number 1–2 and member 2 is defined as node number 2–3. This order needs to be used when writing the $[K]$ matrix using the action and displacement vector using Equations (9.2) and (9.3), respectively. Note that it could have been defined differently – for example, member 1 could have been defined as 2–1.

The KI of this structure is 3 and the unknown displacements are D_1, D_2, and D_3 (translation at joint 1 and rotations at joints 2 and 3). Figure 9.3 shows that the displacement numbering is done starting with the free displacements followed by the restrained displacements. In this problem, D_1, D_2, and D_3 are free displacements (unknowns), while D_4 through D_6 are the restrained displacements, which are zero in this problem. As explained earlier, the actions $\{A_F\}$ corresponding to $\{D_F\}$ are known (which can easily be obtained from the fixed end moments). The $\{A_R\}$ corresponding to the known $\{D_R\}$ are the unknowns.

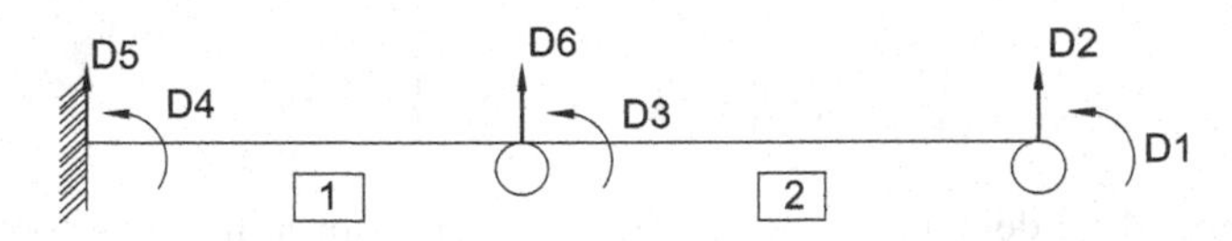

FIGURE 9.3 Beam showing displacements.

Step 4

The stiffness matrix $[K_1]$ for member 1 is obtained from equation (9.2) using the given properties of the beam element in the given problem (Table 9.1).

The above matrix, $[K_1]$, relates the action vector $\{A_5, A_4, A_6, A_3\}^t$ to the displacement vector $\{D_5, D_4, D_6, D_3\}^t$. As pointed out earlier in the basic procedure, displacements numbering follows the order of translation followed by rotations. The same order is followed by the corresponding actions as well. This is very important as noted earlier.

TABLE 9.1
Stiffness matrix for member 1

	5	4	6	3	
$[K_1] =$	89.82	5389.17	-89.82	5389.17	5
	5389.17	431133.33	-5389.17	215566.67	4
	-89.82	-5389.17	89.82	-5389.17	6
	5389.17	215566.67	-5389.17	431133.33	3

A derivation of some terms used in calculating the stiffness matrix $[K]$ is shown below.

$$K_{55} = K_{66} = -K_{65} = -K_{56} = \frac{12EI}{L^3} = \frac{12(29000)(446)}{(10*12)^3} = 89.82$$

Similarly, other terms in the stiffness matrix $[K_1]$ can be calculated.
In exactly the same way, $[K_2]$, the stiffness matrix for member 2 can be assembled using equations (9.2) and (9.3) (as given in Table 9.3).

The above $[K]$ matrix relates the action vector $\{A_6, A_3, A_2, A_1\}^t$ to the displacement vector $\{D_6, D_3, D_2, D_1\}^t$. Again, translations are numbered first followed by rotations.

Step 5

The structure stiffness matrix for this problem $[K_J]$ can be assembled using equation (9.4) and the assembled matrices $[K_1]$ and $[K_2]$ (Table 9.3).

From the given condition, we can calculate

$$\{A_{FN}\} = \begin{Bmatrix} 0 \\ -36 \\ 0 \end{Bmatrix} \begin{matrix} 4 \\ 5 \\ 6 \end{matrix}$$

From the above, the $[K_{FF}]$ can be written as

TABLE 9.2
Stiffness matrix for member 2

	6	3	2	1	
$[K_2] =$	51.98	3742.48	-51.98	3742.48	6
	3742.48	359277.78	-3742.48	179638.89	3
	-51.98	-3742.48	51.98	-3742.48	2
	3742.48	179638.89	-3742.48	359277.78	1

TABLE 9.3
Global/structure stiffness matrix

$[K_J] =$

	1	2	3	4	5	6	
	359277.8	-3742.5	179638.9	0	0	3742.5	1
	-3742.5	51.98	-3742.5	0	0	-51.98	2
	179638.9	-3742.5	790411.1	215566.7	5389.17	-1646.7	3
	0	0	215566.7	431133.3	5389.17	-5389.2	4
	0	0	5389.2	5389.2	89.82	-89.82	5
	3742.5	-51.98	-1646.7	-5389.2	-89.82	141.8	6

$$\left[K_{FF}\right]=\begin{bmatrix} 359277.8 & -3742.5 & 179638.9 \\ -3742.5 & 51.98 & -3742.5 \\ 179638.9 & -3742.5 & 790411.1 \end{bmatrix}$$

We can solve the displacement:

$$\left\{D_F\right\}=\left[K_{FF}\right]^{-1}*\left\{A_{FN}\right\}=\begin{bmatrix} -0.04088 \\ -4.50183 \\ -0.01202 \end{bmatrix}$$

Calculate the reaction forces:

$$\begin{bmatrix} A_4 \\ A_5 \\ A_6 \end{bmatrix}=\left[K_{RF}\right]\left\{D_F\right\}=\begin{bmatrix} -2592^{k-in} \\ -64.8k \\ 100.8k \end{bmatrix}=\begin{bmatrix} -216^{k-ft} \\ -64.8k \\ 100.8k \end{bmatrix}$$

9.6 SUMMARY

This chapter described the basic concepts of one of the most powerful methods of structural analysis – the direct stiffness method. A general procedure for solution of problems by the direct stiffness method is described with application to continuous beams. The method as described in this chapter is general enough so that it can be applied to truss and frame as well.

PROBLEMS FOR CHAPTER 9

Analyze Problem 6.1 in Chapter 6 and Problem 7.1 in Chapter 7 using the direct stiffness method.

Problem 9.1 Solve *Problem 6.1* using the direct stiffness method.

Problem 6.1 repeated: Determine the reactions at the supports of the beam shown in this figure. Take EI as constant.

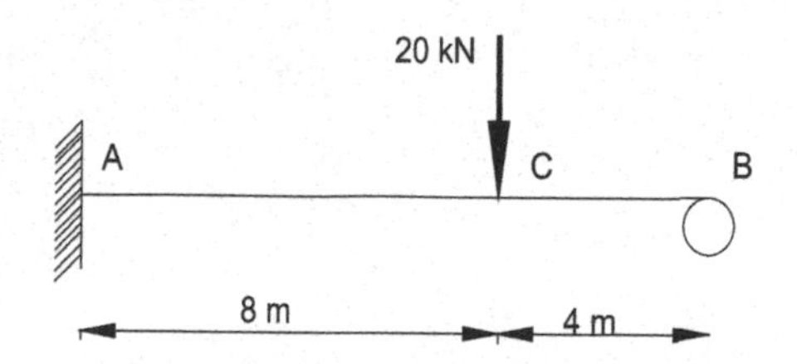

Problem 9.2 Solve *Problem 7.1* using the direct stiffness method.

Problem 7.1 repeated Determine the reactions at the supports of the beam shown in this figure. EI is constant.

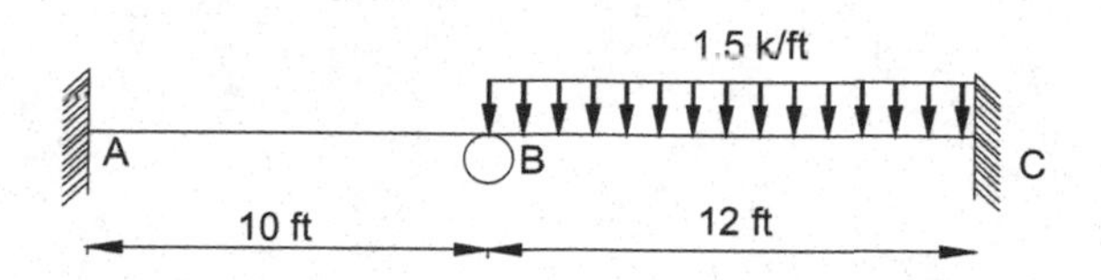

10 Direct Stiffness Method
Application to Frames

10.1 DERIVATION/EXPLANATION OF THE STIFFNESS MATRIX FOR A FRAME ELEMENT

The basic procedure of the direct stiffness method has been explained in Chapter 9. The explanation is essentially a general procedure, which is also applicable to frames. The only difference is that in the case of a frame, the element stiffness matrix $[K]$ has to be used for column and beam elements of a frame (take out rather than for a beam element). This is explained below.

Consider the general frame element shown in Figure 10.1.

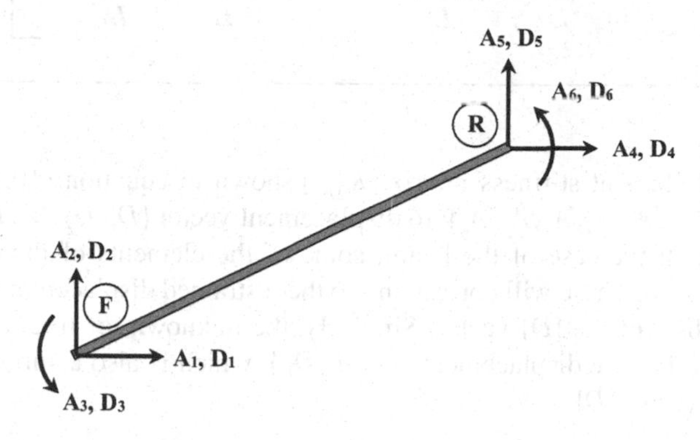

FIGURE 10.1 Typical frame element with free and restrained supports.

The member stiffness matrix for the frame element in structural (global) coordinates, $[K_{MSi}]$, is given as follows:

DOI: 10.1201/9781003246633-12

EQUATION 10.1
Frame element stiffness matrix

$$
\left[K_{MSi}\right]=
\begin{array}{c}
\begin{array}{ccccccc}
1 & 2 & 3 & 4 & 5 & 6 & \\
\end{array}\\
\left[\begin{array}{ccccccc}
\frac{AE}{L} & 0 & 0 & -\frac{AE}{L} & 0 & 0 & 1\\
0 & \frac{12EI}{L^3} & \frac{6EI}{L^2} & 0 & -\frac{12EI}{L^3} & \frac{6EI}{L^2} & 2\\
0 & \frac{6EI}{L^2} & \frac{4EI}{L} & 0 & -\frac{6EI}{L^2} & \frac{2EI}{L} & 3\\
-\frac{AE}{L} & 0 & 0 & \frac{AE}{L} & 0 & 0 & 4\\
0 & -\frac{12EI}{L^3} & -\frac{6EI}{L^2} & 0 & \frac{12EI}{L^3} & -\frac{6EI}{L^2} & 5\\
0 & \frac{6EI}{L^2} & \frac{2EI}{L} & 0 & -\frac{6EI}{L^2} & \frac{4EI}{L} & 6
\end{array}\right]
\end{array}
\qquad (10.1)
$$

The global element stiffness matrix $[K_{MSi}]$ shown in equation (10.1) relates the action vector $\{A_1, A_2, A_3, A_4, A_5, A_6\}^t$ to displacement vector $\{D_1, D_2, D_3, D_4, D_5, D_6\}^t$.

Again, like in the case of the beam, some of the elements of the displacement vector will be zero. These will correspond to the restrained displacement vector $\{D_R\}$, which is a subset of the $\{D\}$ vector. Similarly, the unknown, or free, displacements will be part of the free displacement vector $\{D_F\}$, which is also a subset of the total displacement vector $\{D\}$.

10.2 APPLICATION OF THE DIRECT STIFFNESS METHOD TO A FRAME

The process used to solve for the reactions of an indeterminate frame using the direct stiffness method is explained below using Example 10.2.1.

EXAMPLE 10.2.1

Determine the reactions at the supports of the frame shown in Figure 10.2 using the direct stiffness method. $A = 100$ in^2, $E = 29{,}000$ ksi, and $I = 833$ in^4.

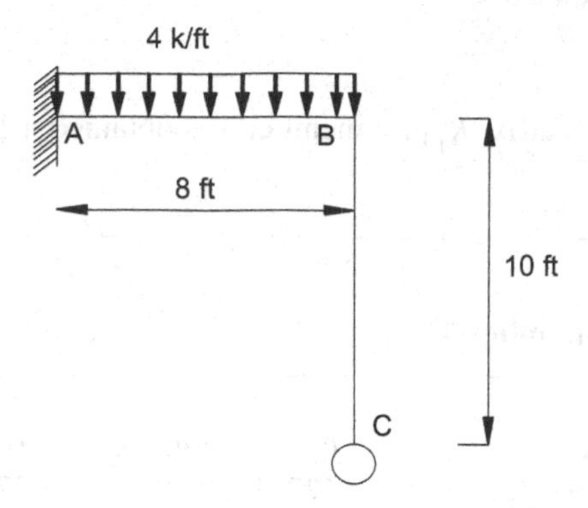

FIGURE 10.2 Indeterminate frame.

Solution

Before starting the solution of the problem, it is to be noted that this frame has side sway (joint translation). However, an important point is that in the case of stiffness method, the procedure of analysis of a frame with or without joint translation is the same except that when the frame has side sway, the kinematic indeterminacy (KI) increases. This will, in essence, involve more equations to be solved.

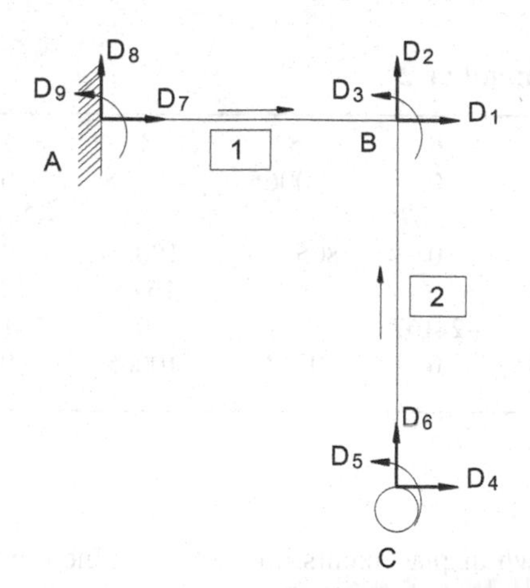

FIGURE 10.3 Frame showing displacements.

Steps 1–3

The numbering process is shown in Figure 10.3. As shown in the figure, KI of this structure is 5. So, the $\{D_F\}$ vector is $\{D_1, D_2, D_3, D_4, D_5\}$. In this problem again, as in the beam problem, the numbering of the displacements is done from the free displacements. The restrained displacement vector, $\{D_R\}$, for this problem is $\{D_6, D_7, D_8, D_9\}$. Only the free displacements are to be obtained, from $[K_{FF}]$ and $\{A_{FN}\}$, as a solution to the problem. Then, the forces (actions) corresponding to the $\{D_R\}$ vector need to be calculated.

Step 4

The member stiffness matrix $[K_1]$ for member 1 is obtained using equation (10.1). It is given as (Table 10.1)

TABLE 10.1
Stiffness matrix for member 1

	7	8	9	1	2	3	
	30208	0	0	–30208	0	0	7
	0	327.7	15727	0	–327.7	15727	8
$[K_1] =$	0	15727	1006542	0	–15727	503271	9
	–30208	0	0	30208	0	0	1
	0	–327.7	–15727	0	327.7	–15727	2
	0	15727	503271	0	–15727	1006542	3

Similarly, the stiffness matrix for member 2 $[K_2]$ is given as (Table 10.2)

TABLE 10.2
Stiffness matrix for member 2

	4	6	5	1	2	3	
	167.8	0	–10065	–167.8	0	–10065	4
	0	24167	0	0	–24167	0	6
$[K_2] =$	–10065	0	805233	10065	0	402617	5
	–167.8	0	10065	167.8	0	10065	1
	0	–24167	0	0	24167	0	2
	–10065	0	402617	10065	0	805233	3

Step 5

Since only the unknown displacements are to be obtained in this problem, we need only $[K_{FF}]$ and $\{A_{FN}\}$ corresponding to the free displacements.

Here, $[K_{FF}]$ can be written as (Table 10.3)

TABLE 10.3
Global/structure stiffness matrix of the frame

$$[K_J] = \left[\begin{array}{ccccc|cccc} 30376.1 & 0 & 10065.4 & -167.8 & 10065.4 & 0 & -30208.3 & 0 & 0 \\ 0 & 24494.3 & -15727 & 0 & 0 & -24166.7 & 0 & -327.7 & -15727 \\ 10065.4 & -15727.2 & 1811775 & -10065.4 & 402617 & 0 & 0 & 15727 & 503271 \\ -167.8 & 0 & -10065.4 & 167.8 & -10065.4 & 0 & 0 & 0 & 0 \\ 10065.4 & 0 & 402617 & -10065.4 & 805233 & 0 & 0 & 0 & 0 \\ \hline 0 & -24166.7 & 0 & 0 & 0 & 24166.7 & 0 & 0 & 0 \\ -30208.3 & 0 & 0 & 0 & 0 & 0 & 30208.3 & 0 & 0 \\ 0 & -327.7 & 15727 & 0 & 0 & 0 & 0 & 327.7 & 15727 \\ 0 & -15727.2 & 503271 & 0 & 0 & 0 & 0 & 15727 & 1006542 \end{array}\right]$$

	1	2	3	4	5	6	7	8	9
1	30376.1	0	10065.4	–167.8	10065.4	0	–30208.3	0	0
2	0	24494.3	–15727	0	0	–24166.7	0	–327.7	–15727
3	10065.4	–15727.2	1811775	–10065.4	402617	0	0	15727	503271
4	–167.8	0	–10065.4	167.8	–10065.4	0	0	0	0
5	10065.4	0	402617	–10065.4	805233	0	0	0	0
6	0	–24166.7	0	0	0	24166.7	0	0	0
7	–30208.3	0	0	0	0	0	30208.3	0	0
8	0	–327.7	15727	0	0	0	0	327.7	15727
9	0	–15727.2	503271	0	0	0	0	15727	1006542

The $\{A_{FN}\}$ can be written as

$$\{A_{FN}\} = \begin{Bmatrix} 0 \\ -16 \\ 256 \\ 0 \\ 0 \end{Bmatrix} \begin{matrix} 1 \\ 2 \\ 3 \\ 4 \\ 5 \end{matrix}$$

In this problem there is a uniformly distributed load on element 1. If there are member loads acting on a particular structure (the way we had in the beam problem in Chapter 9), then fixed end moments have to be calculated and the corresponding $\{A_{FN}\}$ vector can be assembled without any problem. The fixed end moments will be reverse (opposite sign).

Solving the above matrix (which essentially consists of solving five simultaneous equations), all the unknown displacements can be obtained in the following way as:

$$\{D_F\} = [K_{FF}]^{-1} * \{A_{FN}\} = \begin{bmatrix} 0 \\ -4.95x10^{-4} \\ 2.47x10^{-4} \\ 0.02959 \\ 2.47x10^{-4} \end{bmatrix}$$

Calculate the reaction forces

$$\begin{bmatrix} A_6 \\ A_7 \\ A_8 \\ A_9 \end{bmatrix} = [K_{RF}]\{D_F\} + \begin{bmatrix} 0 \\ 0 \\ 16k \\ 256^{k-in} \end{bmatrix} = \begin{bmatrix} 11.96k \\ 0 \\ 20.04\ k \\ 387.9^{k-in} \end{bmatrix} = \begin{bmatrix} 11.96\boldsymbol{k} \\ 0 \\ 20.04\ \boldsymbol{k} \\ 32.325^{k-ft} \end{bmatrix}$$

This completes the solution of the given problem.

10.3 SUMMARY

In this chapter, direct stiffness method was applied to a frame problem. The simplicity of the direct stiffness method, as applied to frames, is that no distinction is to be made for frames with or without joint movements. The only difference is that the problem has to be solved for additional unknowns, but the procedure remains the same.

PROBLEMS FOR CHAPTER 10

Analyze Problem 6.2 in Chapter 6 and Problems 7.3 and 7.4 in Chapter 7 using the direct stiffness method.

Problem 10.1 Solve *Problem 6.2* using the direct stiffness method.

Problem 6.2 repeated: Determine the reactions at the supports of the frame shown in this figure. EI is constant.

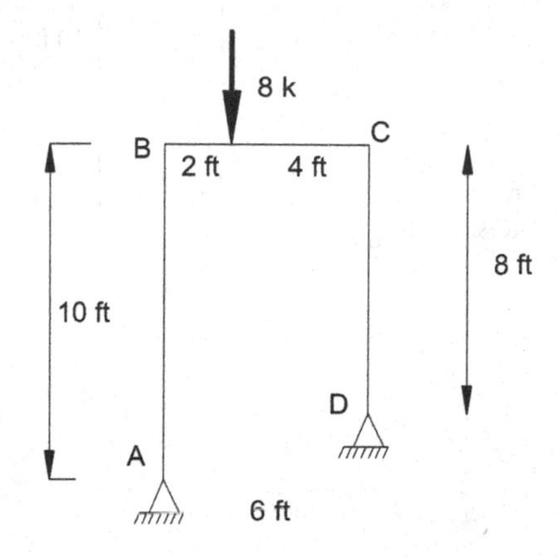

Problem 10.2 Solve P*roblem 7.2* using the direct stiffness method.

Problem 7.2 repeated: Determine the reactions at the supports of the frame shown in this figure. $A = 100$ in^2, $E = 29{,}000$ ksi, and $I = 833$ in^4.

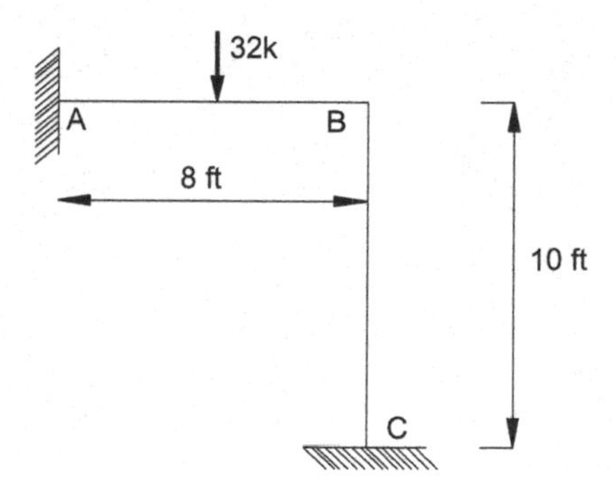

Problem 10.3 Solve *Problem 7.3* using the direct stiffness method.

Problem 7.3 repeated: Determine the reactions at the supports of the frame shown in this figure. $A = 100$ in^2, $E = 29{,}000$ ksi, and $I = 833$ in^4.

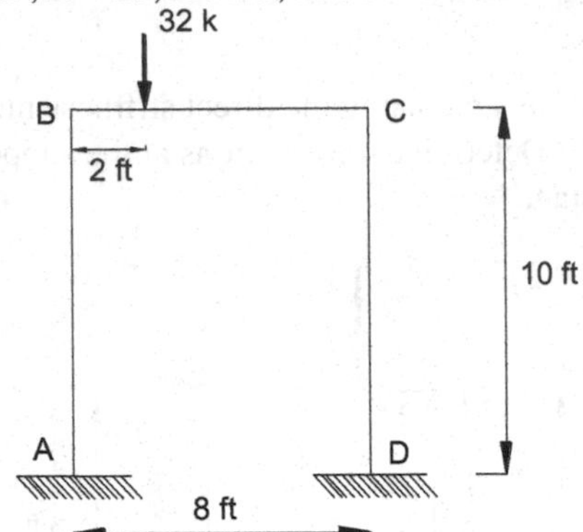

11 Direct Stiffness Method
Application to Trusses

11.1 DERIVATION/EXPLANATION OF THE STIFFNESS MATRIX FOR A TRUSS ELEMENT

As stated earlier, the basic procedure of direct stiffness method has been explained in Chapter 4. The explanation is essentially a general procedure, which is also applicable to trusses. The only difference is that in the case of a truss, the element stiffness matrix $[K]$ for a truss element has to be used instead of that for a beam element. It should also be noted that the members of a truss are subjected to tension or compressive forces only as all the loads on the truss are nodal loads and not member loads. This means that the members of the truss are not subjected to any bending. While this doesn't make any difference in the application of the direct stiffness method, it is an important point to be noted. This concept will be clearer once an example of a truss is solved using the direct stiffness method. Consider the general truss element shown in Figure 11.1.

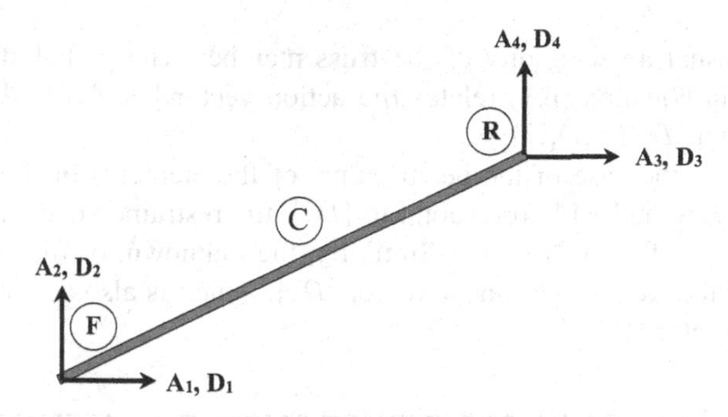

FIGURE 11.1 Typical truss element with free and restrained supports.

The member stiffness matrix for the truss element in structural (global) coordinates, $\left[K_{MSi}\right]$, is given in equation (11.1) as

DOI: 10.1201/9781003246633-13

EQUATION 11.1
Truss element stiffness matrix

$$[K_{MSi}] = \frac{AE}{L} \begin{bmatrix} & D_1 & D_2 & D_3 & D_4 & \\ & C_x^2 & C_xC_y & -C_x^2 & -C_xC_y & A_1 \\ & C_xC_y & C_y^2 & -C_xC_y & -C_y^2 & A_2 \\ & -C_x^2 & -C_xC_y & C_x^2 & C_xC_y & A_3 \\ & -C_xC_y & -C_y^2 & C_xC_y & C_y^2 & A_4 \end{bmatrix} \quad (11.1)$$

where C_x and C_y are the direction cosines of the members given as (see Figure 11.1 for reference). In the above equation "E" is Young's modulus and "A" is member cross section area.

$$C_x = \frac{(x_k - x_j)}{L} \quad (11.2)$$

$$C_y = \frac{(y_k - y_j)}{L} \quad (11.3)$$

EA is the usual axial rigidity of the truss member. The global stiffness matrix $[K_{MSi}]$ shown in equation (6.1) relates the action vector $\{A_1, A_2, A_3, A_4\}^t$ to the displacement vector $\{D_1, D_2, D_3, D_4\}^t$.

Again, like in the case of the beam, some of the elements of the displacement vector will be zero and will correspond to $\{D_R\}$, the restrained displacement vector, which is a subset of the $\{D\}$ vector. Similarly, the unknown, or free, displacements will be part of the free displacement vector $\{D_F\}$, which is also a subset of the total displacement vector $\{D\}$.

11.2 APPLICATION OF THE DIRECT STIFFNESS METHOD TO A TRUSS

The process used to solve for the reactions of an indeterminate truss using the direct stiffness method is explained below using Example 11.2.1.

Example 11.2.1

Determine the reactions at the support of the truss shown in Figure 11.2 using direct stiffness method. AE is constant.

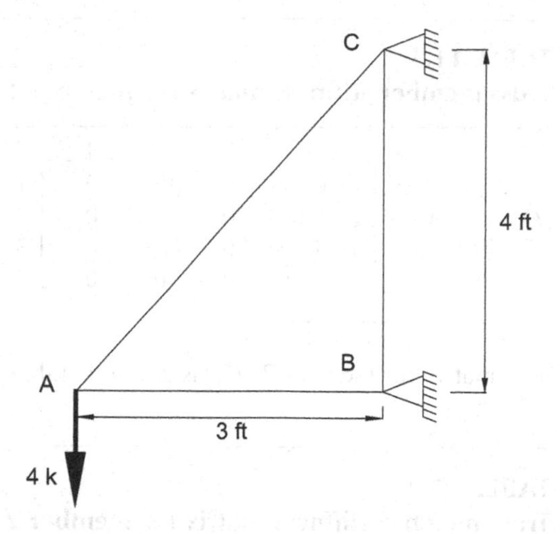

FIGURE 11.2 Indeterminate truss.

Steps 1–3

These steps are shown in Figure 11.3. This figure shows that the kinematic indeterminacy (KI) of this structure is 2. Thus, the $\{D_F\}$ vector is: $\{D_1, D_2\}$. In this problem, just as in the beam and frame problems, the numbering of the displacements is done starting with the free displacements. The restrained displacement vector for this problem, $\{D_R\}$, is $\{D_3, D_4, D_5, D_6, D_7, D_8\}$. Only the free displacements are to be obtained from $[K_{FF}]$ and the $\{A_{FN}\}$ as a solution of the problem.

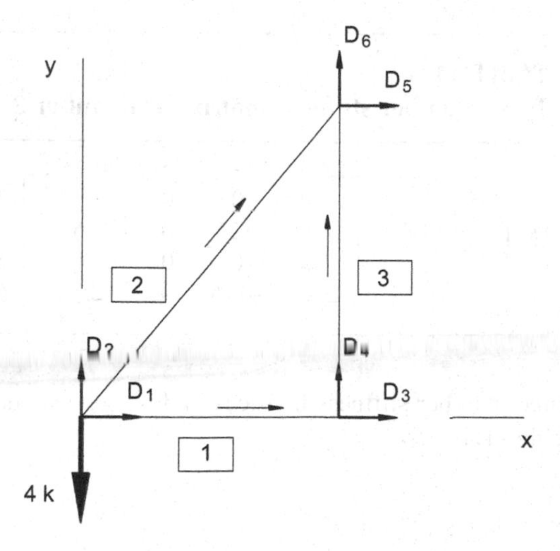

FIGURE 11.3 Truss showing displacements.

Step 4

The member stiffness matrix $[K_1]$ for member 1 is obtained using equation (11.1). It is given in Table 11.1.

TABLE 11.1
Truss member stiffness matrix for member 1

$$[K_1] = AE \begin{bmatrix} 0.333 & 0 & -0.333 & 0 \\ 0 & 0 & 0 & 0 \\ -0.333 & 0 & 0.333 & 0 \\ 0 & 0 & 0 & 0 \end{bmatrix}$$

	1	2	3	4	
$[K_1] = AE$	0.333	0	−0.333	0	1
	0	0	0	0	2
	−0.333	0	0.333	0	3
	0	0	0	0	4

Similarly, the stiffness matrix for member 2 $[K_2]$ is given in Table 11.2.

TABLE 11.2
Truss member stiffness matrix for member 2

	1	2	5	6	
$[K_2] = AE$	0.072	0.096	−0.072	−0.096	1
	0.096	0.128	−0.096	−0.128	2
	−0.072	−0.096	0.072	0.096	5
	−0.096	−0.128	0.096	0.128	6

Finally, the stiffness matrix for member 3 $[K_3]$ is given in Table 11.3.

TABLE 11.3
Truss member stiffness matrix for member 3

	3	4	5	6	
$[K_3] = AE$	0	0	0	0	3
	0	0.25	0	−0.25	4
	0	0	0	0	5
	0	−0.25	0	0.25	6

Combining the three member stiffness matrices yields the structure stiffness matrix given below in Table 11.4.

TABLE 11.4
Truss structure stiffness matrix

$$[K_J] = AE \begin{array}{c} \begin{array}{cccccc} \quad 1 & \quad\; 2 & \quad\;\; 3 & \quad 4 & \quad\; 5 & \quad\;\; 6 \end{array} \\ \begin{bmatrix} 0.405 & 0.096 & -0.333 & 0 & -0.072 & -0.096 \\ 0.096 & 0.128 & 0 & 0 & -0.096 & -0.128 \\ -0.333 & 0 & 0.333 & 0 & 0 & 0 \\ 0 & 0 & 0 & 0.25 & 0 & -0.25 \\ -0.072 & -0.096 & 0 & 0 & 0.072 & 0.096 \\ -0.096 & -0.128 & 0 & -0.25 & 0.096 & 0.378 \end{bmatrix} \end{array} \begin{array}{c} \\ 1 \\ 2 \\ 3 \\ 4 \\ 5 \\ 6 \end{array}$$

Step 5

Since just the unknown displacements are to be obtained in this problem, we only need $[K_{FF}]$ and $\{A_{FN}\}$ corresponding to the free displacements.

From the given condition, we can calculate:

$$\{A_{FN}\} = \begin{Bmatrix} 0 \\ -4 \end{Bmatrix} \begin{matrix} 1 \\ 2 \end{matrix}$$

From $[K]$, we can derive $[K_{FF}]$:

$$\left[K_{FF}\right] = AE \begin{bmatrix} 0.405 & 0.096 \\ 0.096 & 0.128 \end{bmatrix}$$

We can solve the displacement:

$$\{D_F\} = \left[K_{FF}\right]^{-1} * \{A_{FN}\} = \frac{1}{AE} \begin{bmatrix} 9 \\ -38 \end{bmatrix}$$

Calculate the reaction forces:

$$\begin{bmatrix} A_3 \\ A_4 \\ A_5 \\ A_6 \end{bmatrix} = \left[K_{RF}\right]\{D_F\} = \begin{bmatrix} -3\,\boldsymbol{kips} \\ 0 \\ 3\,\boldsymbol{kips} \\ 4\,\boldsymbol{kips} \end{bmatrix}$$

In this problem there is only one nodal load (at joint 4). So, the assembly of $\{A_{FN}\}$ is rather simple. However, if there are other loads at other nodes or if there are some support settlements, the corresponding $\{A_{FN}\}$ vector should be assembled accordingly.

This completes the solution of the problem.

11.3 SUMMARY

In this chapter, direct stiffness method has been applied to a truss problem. The point to be noted about the direct stiffness method is that it can be applied to structures that are mainly in flexure (bending) like beams and frames, as well as structures in tension or compression (like a trusses). The method essentially is the same except that the truss element stiffness matrix is used in analysis of trusses as opposed to a beam or frame element stiffness matrix.

Hence, the direct stiffness method is so powerful and popular that it is used for almost all the computer codes prevalent in structural analysis.

PROBLEMS FOR CHAPTER 11

Analyze Problem 6.3 in Chapter 6 using the direct stiffness method.

Problem 11.1 Solve *Problem 6.3* using the direct stiffness method.

Problem 6.3 repeated: Determine the reactions at the supports of the truss shown in this figure. AE is constant.

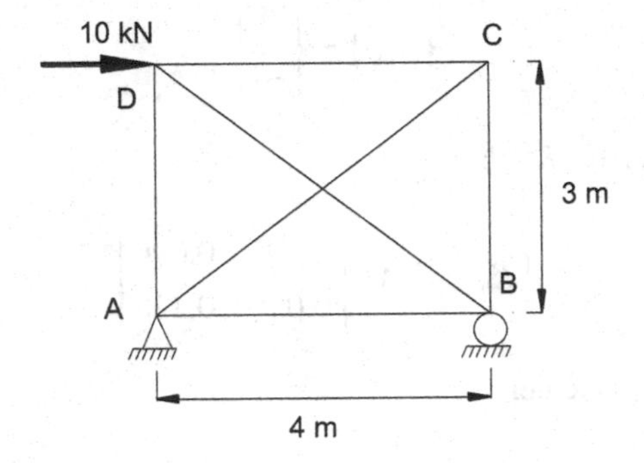

12 Approximate Methods

From a broad viewpoint, analysis of every structure is approximate within the scope of the assumptions needed for the analysis.

12.1 IMPORTANCE OF APPROXIMATE METHODS

1. Analyst's lack of knowledge to carry out the analysis of a highly indeterminate structure.
2. Time required for the analysis may be uneconomical.
3. Preliminary design consists of approximate analysis to find approximate members properties such as I, A, etc.

12.2 ANALYSIS OF A PORTAL

If the frame is symmetric and bending stiffnesses of legs are equal, the solutions from exact analysis show that the horizontal shear on portal will be divided equally between the two legs.

Then, the remaining analysis is carried out as a determinate structure.

If the supports of a portal are fixed, the order of indeterminacy is 3. It will be assumed that horizontal reactions for two legs are equal for symmetric portal. From the deflected shape, we note that the point of inflection occurs at the center of the column. Thus, we provide two hinges, one on each column at their center height.

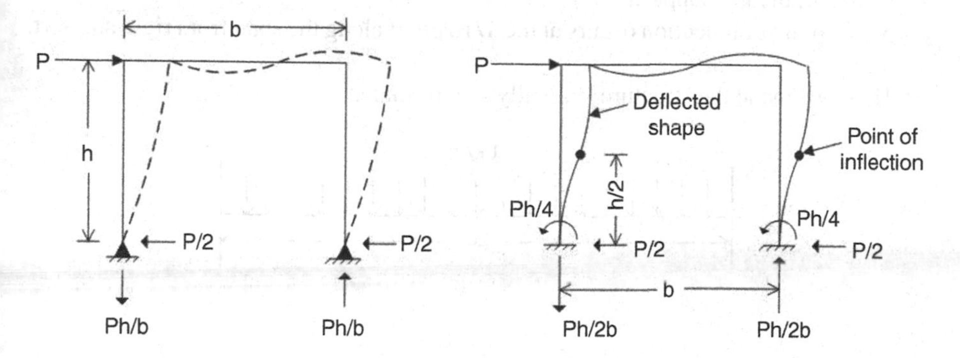

DOI: 10.1201/9781003246633-14

12.3 BUILDING FRAMES UNDER VERTICAL LOADS

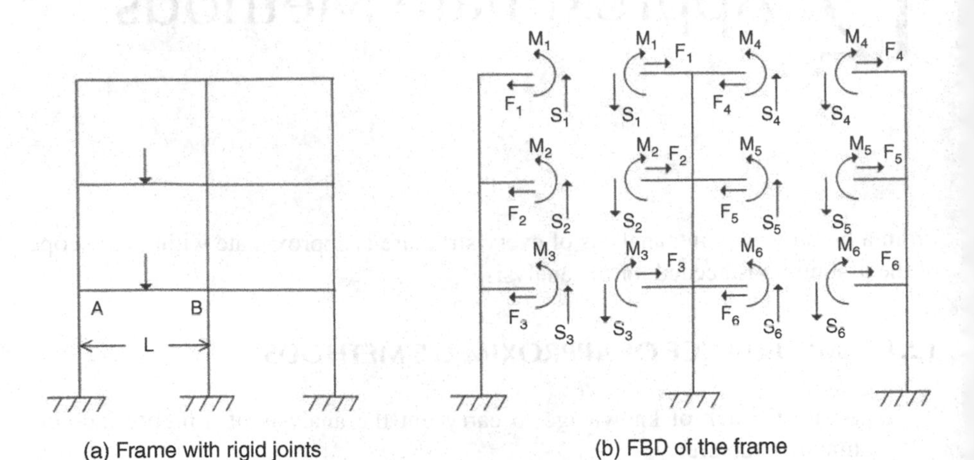

(a) Frame with rigid joints (b) FBD of the frame

If we can determine $F_1, F_2, \ldots, F_6$; $S_1, S_2, \ldots, S_6$; and $M_1, M_2, \ldots, M_6$, then the structure is determinate. But how? Assuming that AB is simply supported at both ends, that is, A and B, then no point of inflection exists between A and B.

Suppose A and B are completely fixed, the point of inflection occurs at $0.21L$ from each support.

However, A and B are neither simple supports nor fixed. Therefore, the point of inflection should fall between 0 and $0.21L$. From all observations, for the approximate analysis the point of inflection is assumed to fall at $0.1L$ from each support.

The following assumptions are made for *the analysis of a beam under vertical loads:*

1. Axial force in girders is zero.
2. A point of inflection occurs at the one-tenth point measured along the span from the left support.
3. A point of inflection occurs at the 1/10 point along the span from right support.

Thus, we made the structure statically determinate.

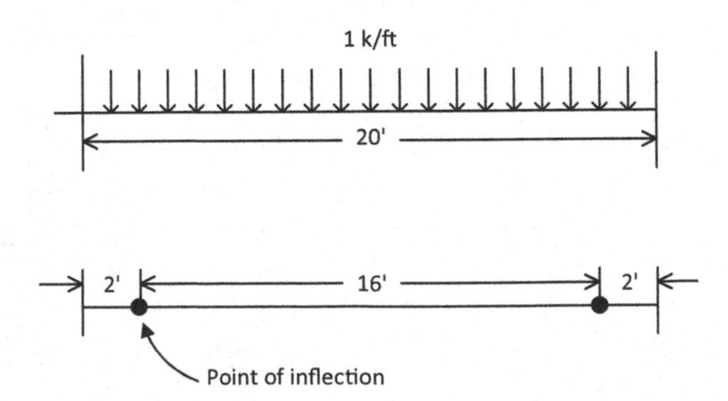

$$\text{Max. Positive Moment} = \frac{wl^2}{8} = \frac{1(16)^2}{8} = 32 \text{ k-ft.}$$

$$\text{Max. Negative Moment} = 8(2) + 2 \times 1 \times \frac{2}{2} = 18 \text{ k-ft.}$$

End shears on girders = vertical force applied to column by girders

Axial column force = Sum of all the end shears from top and bottom.

Maximum axial load on a column is obtained by loading bays on both sides of the column. For interior columns, column moments are neglected because moments acting on both sides of the column are almost equal and opposite in direction.

However, for exterior columns, column moments are considered because they act only on one side. Girder moments are divided between columns in proportion to their stiffnesses.

12.4 BUILDING FRAMES UNDER LATERAL LOADS

Assumptions for analyzing building frames under lateral loads should be different from those frames under vertical loads.

For lateral loads, points of inflection occur near the center of each girder and each column.

12.5 PORTAL METHOD

Assumptions:

1. Points of inflection exist at center of each girder and each column.
2. Total horizontal shear on each story is divided between columns of that story in a manner such that each interior column carries twice as much shear as each exterior one.

These assumptions generally lead to more conditional equations than what is needed. However, additional equations do not cause any inconsistency in results.

12.5.1 Column Shears

First story: x = shear in exterior column, 2x = shear in interior column

$x + 2x + 2x + x = 6x = 10{,}000 + 10{,}000 = 20{,}000$

$\therefore$ $x = 3{,}333$ $2x = 6{,}667$

Second story: $6x = 10{,}000$

$\therefore$ $x = 1{,}667$ $2x = 3{,}333$

Column Moments: Moments at center of each column are zero.

End moment for a column = (Shear on the column)(Half the height)

$M_{AE} = 3333(10 \text{ ft}) = 33{,}333$ ft-lb

$M_{FJ} = 3333(7.5 \text{ ft}) = 25{,}000$ ft-lb

Girder Moments: Girder and column moments act in opposite directions on a joint.

$M_{c1} + M_{c2} = M_{g1} + M_{g2}$

∴ Sum of Column End Moments = Sum of Girder End Moments.

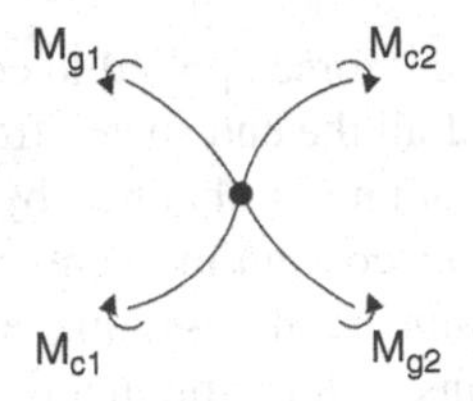

Say at joint E:

$M_{EF} = 33{,}333 + 12{,}500 = 45{,}833$ ft-lb
$M_{FE} = 45{,}833$ because of point of inflection at the center of the girder

At Joint F:

$M_{FG} + 45{,}833 = 66{,}667 + 25{,}000$
$M_{FG} = 45{,}833$ ft-lb

Girder moments in a roof may be determined in a similar manner each of which is found as 12,500 ft-lb.

12.5.2 Girder Shears

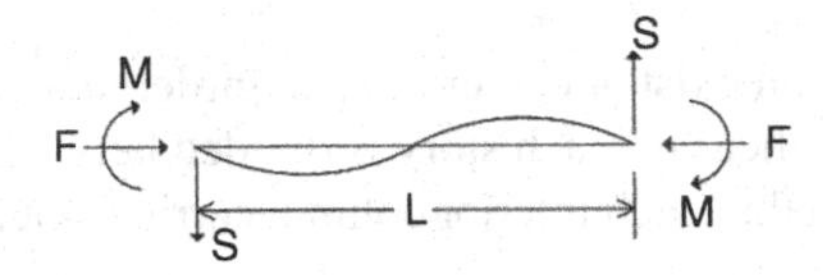

M = 0 for a girder

$$SL = 2M \text{ or } S = \frac{2M}{L}$$

$$\therefore\ S_{EF} = \frac{2 \times 45833}{20} = 4583\ \ ib\ \ or\ \ S_{IJ} = \frac{2 \times 12500}{20} = 1250\ \ ib$$

12.5.3 Column Axial Force

10,000 → I — S_{IJ}=1,250
10,000 → E — S_{EF}=4,583

$F_{EI} = 1250\ lb$, $F_{AE} = 1250 + 4583 = 5833\ lb$

12.6 CANTILEVER METHOD

Assumption:

1. Point of inflection is assumed at the center of each girder and each column.
2. The intensity of axial stress in each column of a story is proportional to the horizontal distance of that column from the center of gravity of all the columns of the story under consideration

If there are '*m*' columns in a story, assumption 2 is equivalent to making $(m-1)$ assumptions regarding column axial-stress relations for each story. Here again, we are making more assumptions than needed. However, the excessive assumptions are consistent with reference to the necessary assumptions.

12.6.1 Column Axial Forces

Assuming all columns have the same cross-sectional area, the C.G. of the columns in each story is

$$x = \frac{20+45+75}{4} = 35\ ft \text{ from AEI}$$

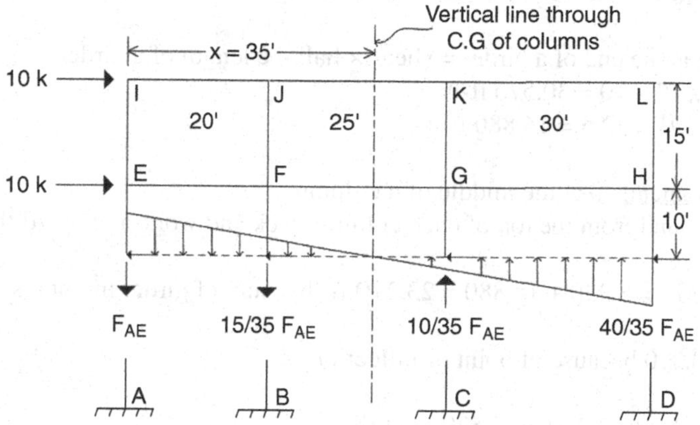

Vertical line through C.G of columns

X = 35'

10 k, 10 k, I, J, K, L, E, F, G, H, 20', 25', 30', 15', 10'

F_{AE} 15/35 F_{AE} 10/35 F_{AE} 40/35 F_{AE}

A B C D

For first story, if the axial force in AE is F_{AE}, by assumption 2, axial force in BF, CG, and DH will be $\frac{15}{35}F_{AE}$, $-\frac{10}{35}F_{AE}$, and $-\frac{40}{35}F_{AE}$. Taking moments about point of inflection of column DH, of forces acting on that part of the bent lying above the horizontal plane passing through points of inflection of the columns of the first story:

$$10{,}000(25)+10{,}000(10)-F_{AE}(75)-\frac{15}{35}F_{AE}(55)+\frac{10}{35}F_{AE}(30)=0$$

$$\therefore F_{AE} = +3890;\ F_{BF} = +1670;\ F_{CG} = -1112;\ F_{DH} = -4500$$

For the second story, the column axial forces are in the same ratio to each other as they are in the first story. They would be evaluated by taking moments at the point of inflection of column HL, that is, center of HL.

$$\therefore F_{EI} = +833;\ F_{FJ} = +358;\ F_{KG} = -238;\ F_{LH} = -953$$

Girder Shears: girder shears are obtained from column axial forces at the various joints.

At joint E, $S_{EF} = +388 - 3890 = -3057$
Similarly, at joint F
$S_{FG} = -3057 + 358 - 1670 = -4369$

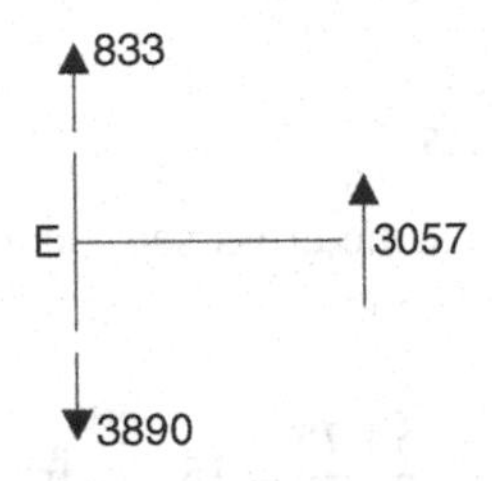

Girder Moments: Moment at the center of each girder = 0

Moment at the end of a girder = shear × half the length of a girder
$M_{EF} = 3{,}057 \times 10 = 30{,}570$ ft-lb
$M_{KJ} = 1{,}191 \times 12.5 = 14{,}880$ ft-lb

Column Moments: 0 at the middle of a column.
These are found from the top of each column stack and working toward the base.

At joint J, $M_{JF} = 8{,}330 + 14{,}880 = 23{,}210$ ft-lb = sum of girder moments

$M_{FJ} = 23{,}210$ because of point of inflection

At joint F, $M_{FB} + 23{,}210 = 30{,}570 + 54{,}600$

$M_{FB} = 61{,}960$ ft-lb $= M_{BF}$ because of inflection point between B and F.

The resulting relative stiffness for each member is shown in Figure 12.1. The resulting bending moments for each member are shown in Figures 12.2–12.4.

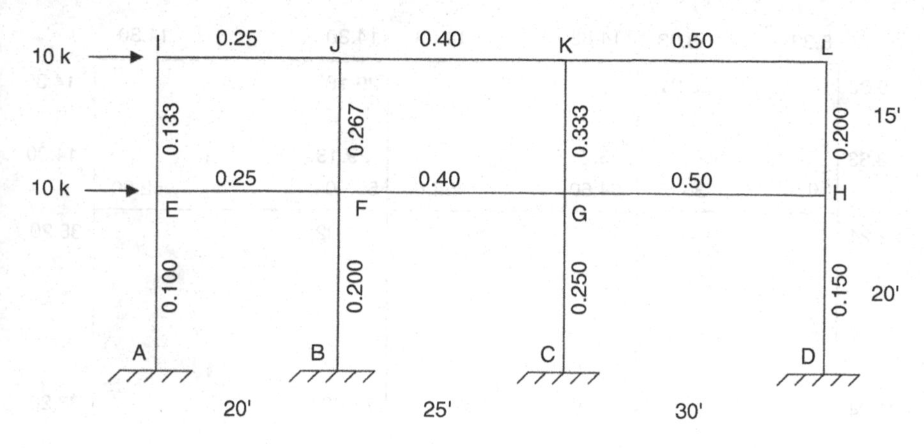

FIGURE 12.1 Relative stiffnesses.

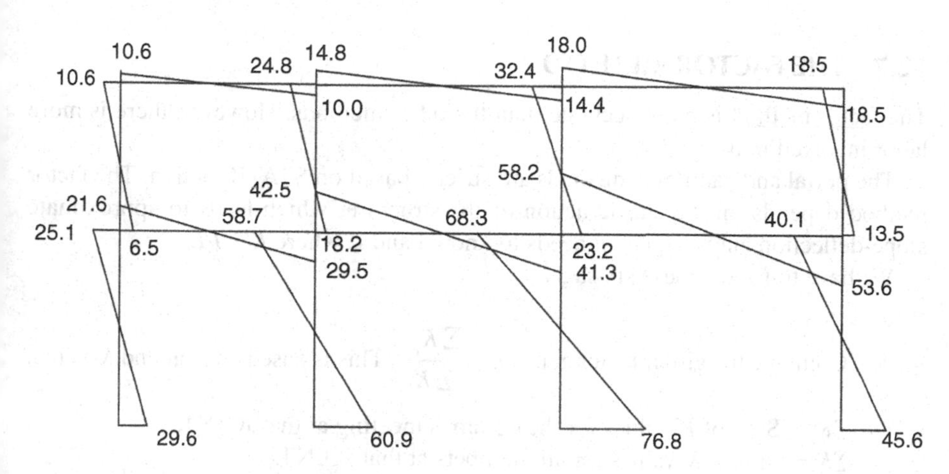

FIGURE 12.2 "Exact" solution.

12.5 12.5 12.5 12.5 12.5 12.5

12.5 25.0 25.0 12.5

12.5 25.0 25.0 12.5

45.8 45.8 45.8 45.8 45.8 45.8

33.3 66.7 66.7 33.3

33.3 66.7 66.7 33.3

FIGURE 12.3 Portal method.

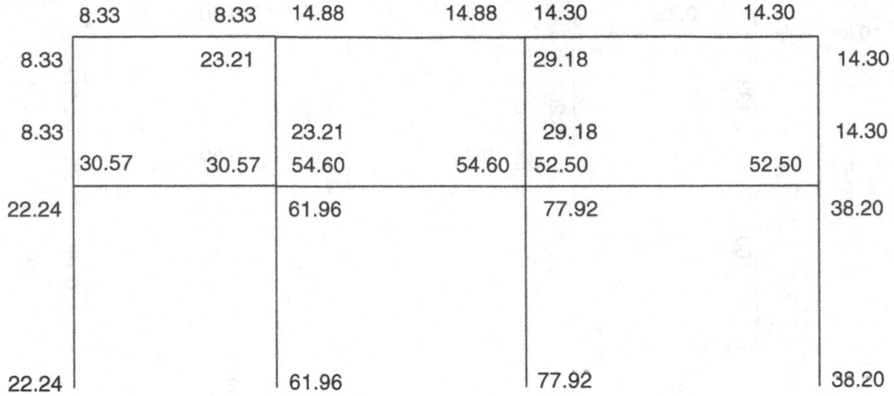

8.33 8.33 14.88 14.88 14.30 14.30
8.33 23.21 29.18 14.30
8.33 23.21 29.18 14.30
30.57 30.57 54.60 54.60 52.50 52.50
22.24 61.96 77.92 38.20
22.24 61.96 77.92 38.20

FIGURE 12.4 Cantilever method.

12.7 THE FACTOR METHOD

The factor method is more accurate than the other methods. However, there is more labor involved in it.

The portal and cantilever methods are strictly based on STATIC action. The factor method depends on the elastic action of the structure, which leads to approximate slope-deflection analysis, i.e., it needs to know I and L where $K = I/L$.

We have to follow these six steps:

1. Compute the girder factor g, i.e., $g = \dfrac{\Sigma K_C}{\Sigma K}$. This is based on near end K-value.

   ΣK_c = Sum of K-values for the columns meeting at that JOINT.
   ΣK = Sum of K-values for all members at that JOINT.

2. Compute the column factor C. Based on near-end K-value

   C = 1-g, where g = girder factor.
   For fixed column bases of first story (supports) C = 1.

3. "g" or "C" at a joint must be added to half the "g" or "C" coming from the FAR end or "other end".
4. Multiply values of step 3 by 'K' value for the member in which the sum occurs. For columns, it is C and for girders it is G.
5. The column moment factors C (of step 4) are the approximate relative values for column end moments for the story in which they occur.

   The sum of column end moments in a given structure = (Total horizontal shear on that story) × Height of story

   Hence, column moment factor C may be converted to column end moments by direct proportion for each story.

6. Girder moment factor G (from step 4) as approximate relative values for girder end moments for each joint.

Sum of girder end moments = Sum of column end moment at that joint (from Step 5).

Hence, G may be converted and girder end moment by direct proportion for each joint.

12.8 MODIFIED PORTAL METHOD

12.8.1 Assumption

1. h = story height
2. L = beam length; inflection point for beam

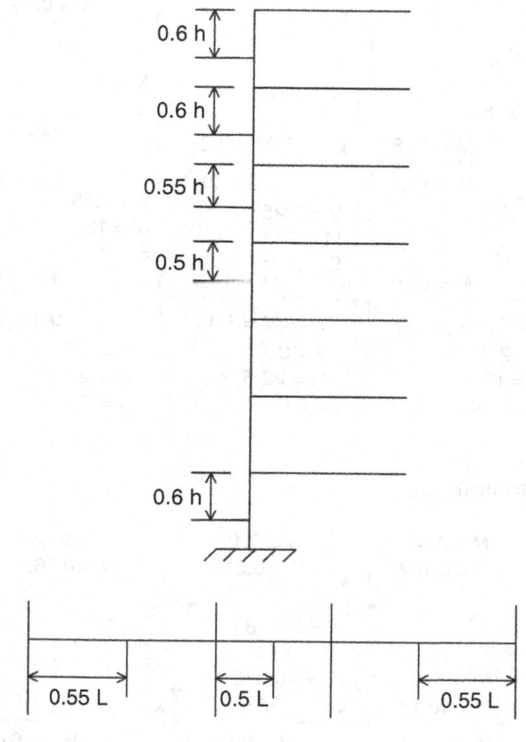

3. Shear split
   (a) Divide among the columns in proportion, i.e., I of each column. That shear is

$$Shear = Story\,Shear\left[\frac{No.of\ Bays - \frac{1}{2}}{No.of\ Columns}\right]$$

For bottom story only

$$Shear = Story\,Shear\left[\frac{No.of\ Bays - 2}{No.of\ Columns}\right]$$

(b) Divide the remaining shear among bays in proportion

$$\left[\frac{I\ of\ beam\ above}{Span\ of\ bay}\right] / \left[\sum_{of\ the\ story} \frac{I\ of\ beam\ above}{Span\ of\ bay}\right]$$

and split equally the bay shear to each adjacent column. The resulting force diagrams for Portal Method (Figure 12.5) and Cantilever Method (Figure 12.6) are shown below.

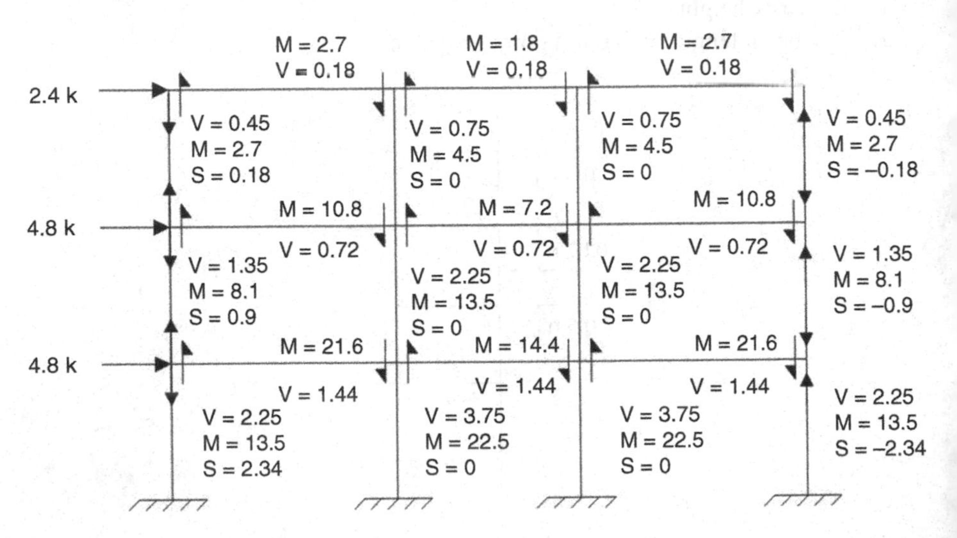

FIGURE 12.5 Portal method.

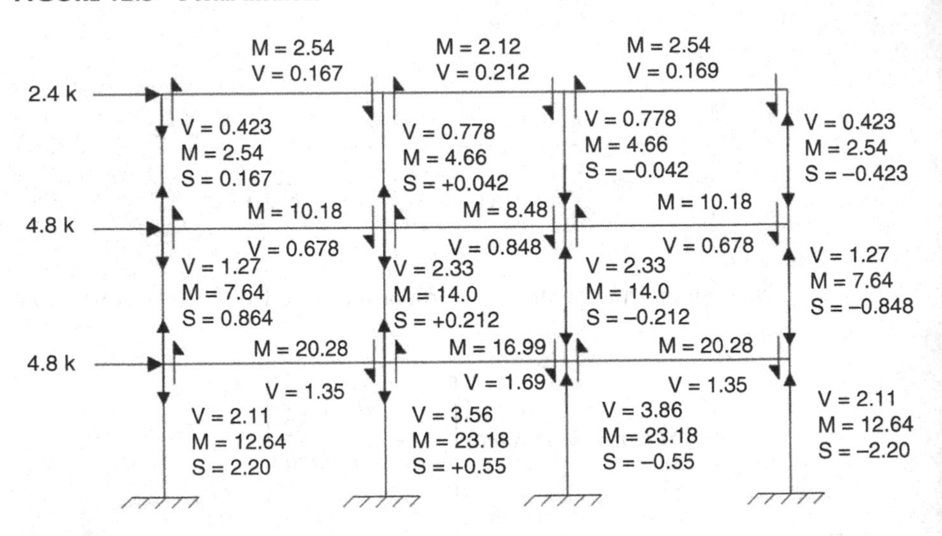

FIGURE 12.6 Cantilever method.

12.9 RESULTS FROM PORTAL METHOD

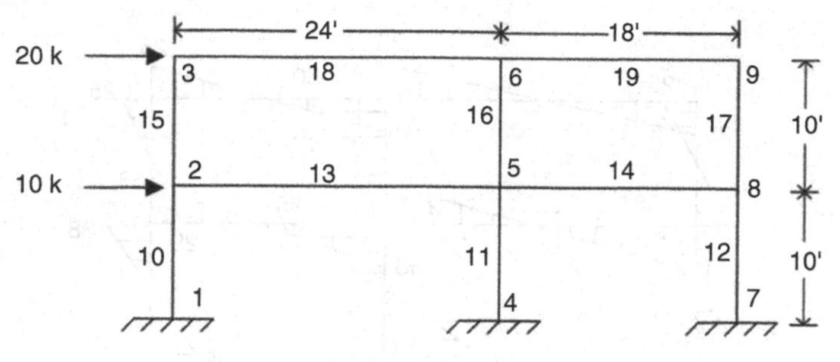

(a) Loads and Layout

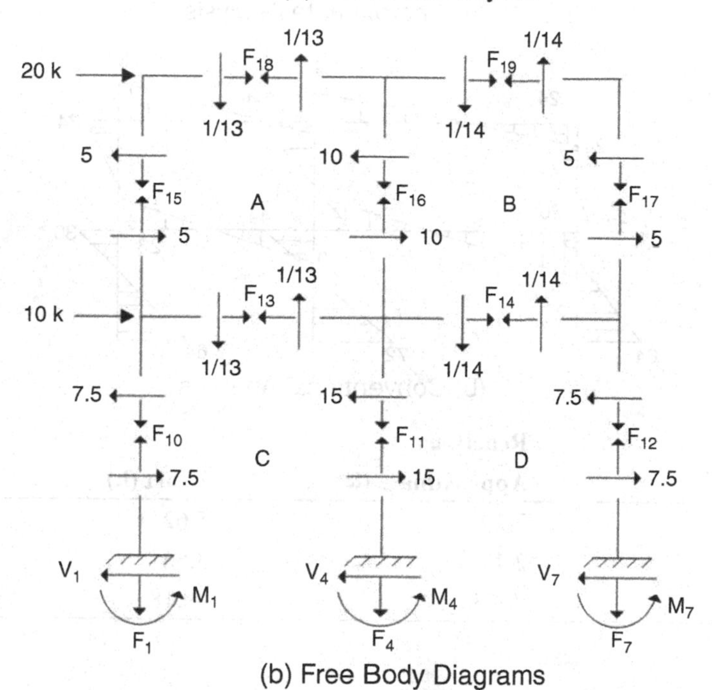

(b) Free Body Diagrams

12.9.1 Results

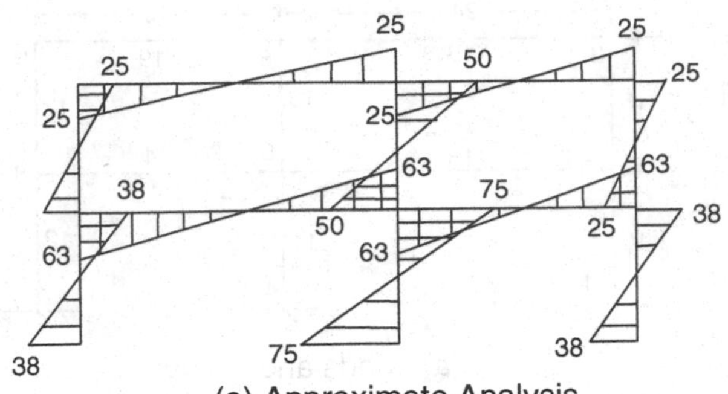

(a) Approximate Analysis

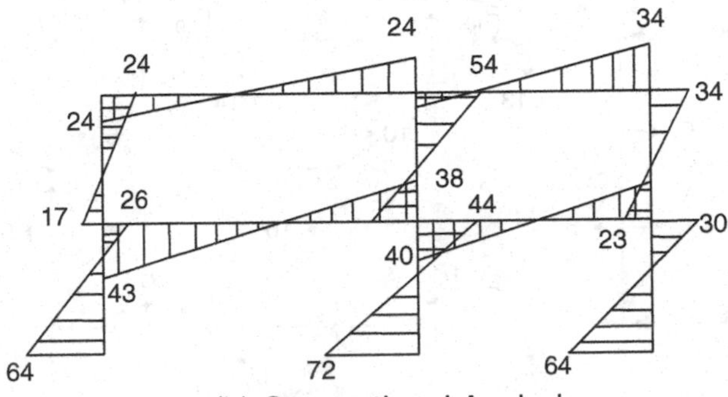

(b) Conventional Analysis

Force	Reactions Approximate (k)	Exact (k)
F_1	7.28	5.62
F_4	2.45	3.56
F_7	-9.73	–9.18

12.9.2 Cantilever Method

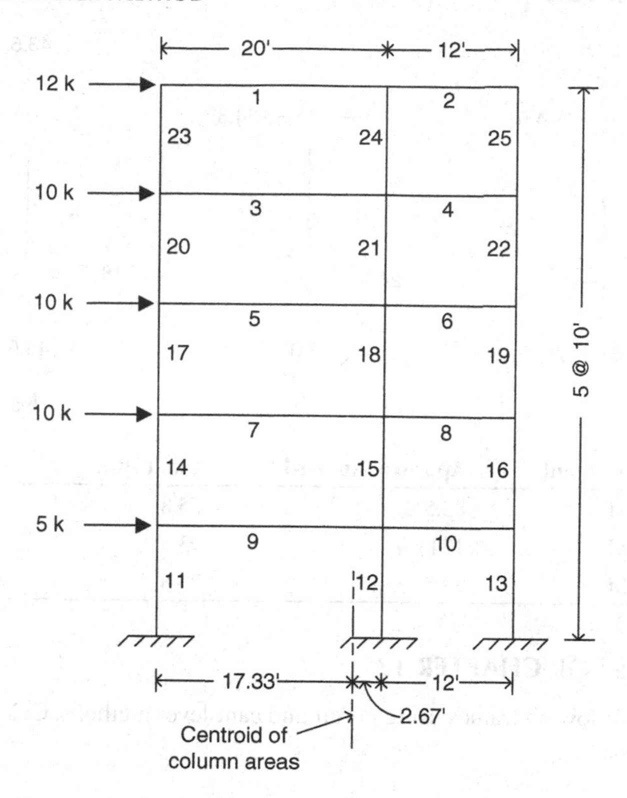

Note: Column forces vary linerarly from the centriod.

12.9.3 Top Floor FBD

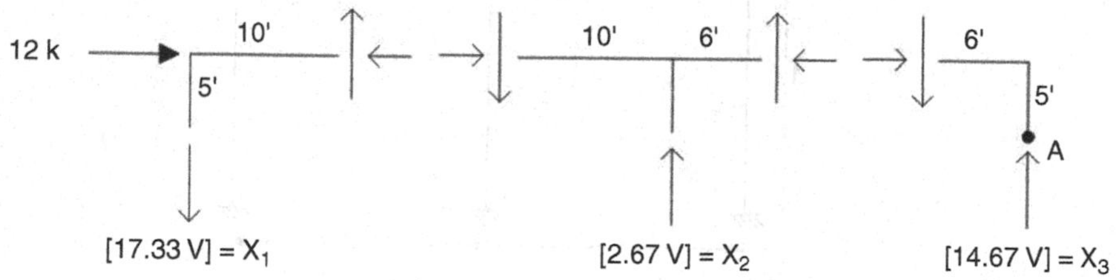

12 k — 10' — 5' — 10' — 6' — 6' — 5' — A

[17.33 V] = X_1 [2.67 V] = X_2 [14.67 V] = X_3

$$\Sigma M_A = 0 = (12)(5) + (12)[2.67V] - (32)[17.33V]$$

V = 0.115

$X_1 = 2.0$

$X_2 = 0.3$

$X_3 = 1.7$

12.9.4 Base FBD

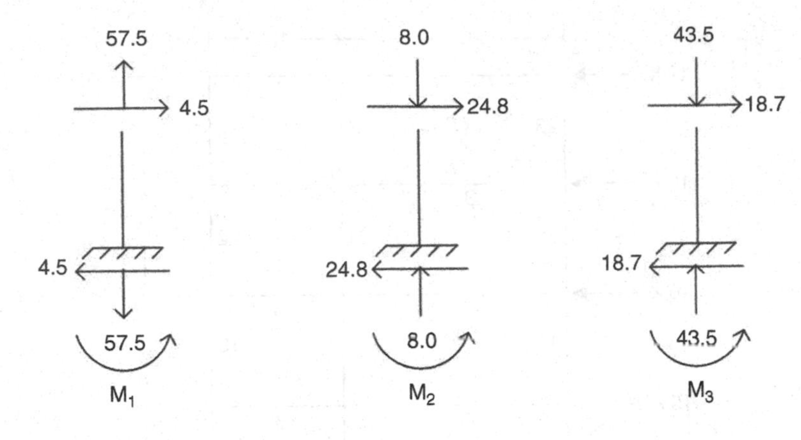

Moment	Approximate (ft-k)	Exact (ft-k)
M_1	82.5	65.8
M_2	124.0	93.5
M_3	28.5	75.8

PROBLEMS FOR CHAPTER 12

Analyze the following frames using portal and cantilever methods. Calculate the support reactions.

Problem 12.1

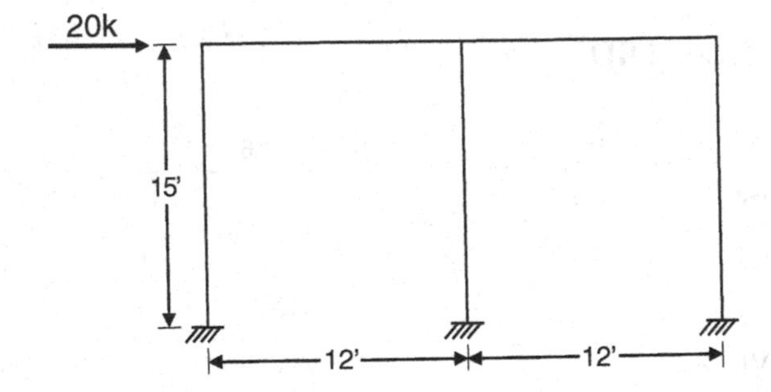

Problem 12.2

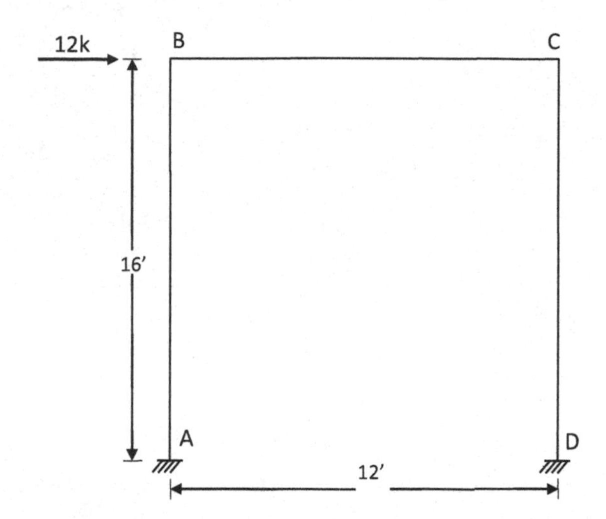

Problem 12.3

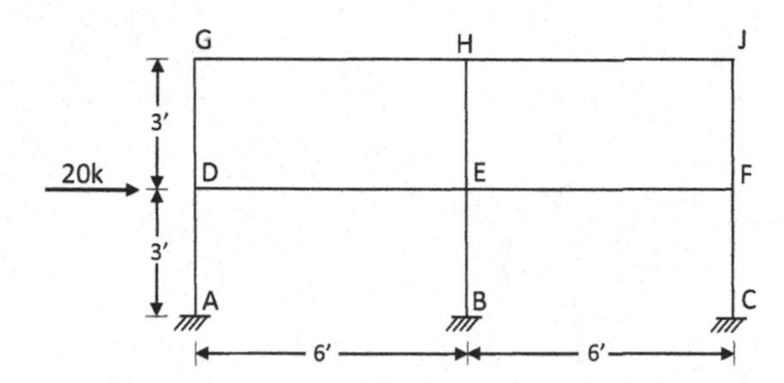

Problem 12.4

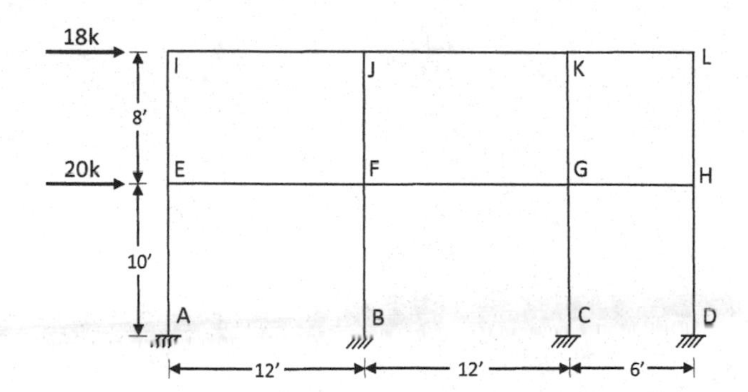

Appendix A

TABLE 1A
Fixed-end moments

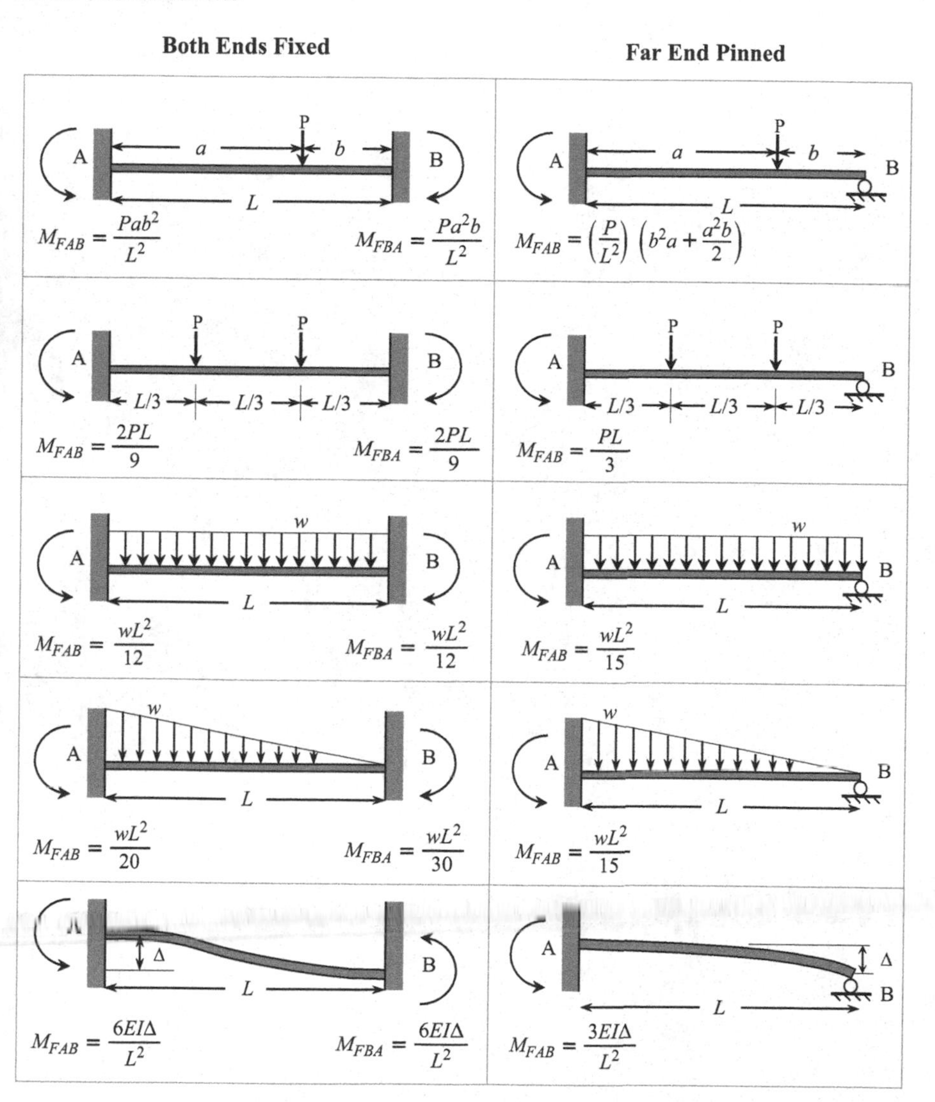

Both Ends Fixed
Far End Pinned
P
a
b
A
B
L
$M_{FAB} = \frac{Pab^2}{L^2}$
$M_{FBA} = \frac{Pa^2b}{L^2}$
$M_{FAB} = \left(\frac{P}{L^2}\right)\left(b^2a + \frac{a^2b}{2}\right)$
L/3
$M_{FAB} = \frac{2PL}{9}$
$M_{FBA} = \frac{2PL}{9}$
$M_{FAB} = \frac{PL}{3}$
w
$M_{FAB} = \frac{wL^2}{12}$
$M_{FBA} = \frac{wL^2}{12}$
$M_{FAB} = \frac{wL^2}{15}$
$M_{FAB} = \frac{wL^2}{20}$
$M_{FBA} = \frac{wL^2}{30}$
$M_{FAB} = \frac{wL^2}{15}$
Δ
$M_{FAB} = \frac{6EI\Delta}{L^2}$
$M_{FBA} = \frac{6EI\Delta}{L^2}$
$M_{FAB} = \frac{3EI\Delta}{L^2}$

References

Chajes, A. (1983). *Structural Analysis*. Prentice-Hall, Inc., Englewood Cliffs, NJ.

Hibbeler, R.C. (2012). *Structural Analysis*. Prentice Hall, Upper Saddle River, NJ.

Wang. C-K. (1953). *Statically Indeterminate Structures*. McGraw-Hill Book Company. New York, NY.

Weaver, W. and Gere, J.M. (1990). *Matrix Analysis of Framed Structures*. McGraw-Hill Van Nostrand Reinhold, New York, NY.

Index

Printed in the United States
by Baker & Taylor Publisher Services

Printed in the United States
by Baker & Taylor Publisher Services